Barcoding Nature

DNA barcoding has been promoted since 2003 as a new, fast, digital genomics-based means of identifying natural species based on the idea that a tiny standard fragment of any organism's genome (a so-called 'micro-genome') can faithfully identify and help to classify every species on the planet. The fear that species are becoming extinct before they have ever been known fuels barcoders, and the speed, scope, economy and 'user-friendliness' claimed for DNA barcoding, as part of the larger ferment around the 'genomics revolution', has also encouraged promises that it could inspire humanity to reverse its biodiversity-destructive habits.

This book is based on six years of ethnographic research on changing practices in the identification and classification of natural species. Informed both by science and technology studies and the anthropology of science, the authors analyse DNA barcoding in the context of a sense of crisis – concerning global biodiversity loss, but also the felt inadequacy of taxonomic science to address such loss. The authors chart the specific changes that this innovation is propelling in the collecting, organizing, analysing and archiving of biological specimens and biodiversity data. As they do so they highlight the many questions, ambiguities and contradictions that accompany the quest to create a genomics-based environmental technoscience dedicated to biodiversity protection. They ask what it might mean to recognize ambiguity, contradiction and excess more publicly as a constitutive part of this and other genomic technosciences.

Barcoding Nature will be of interest to students and scholars of sociology of science, science and technology studies, politics of the environment, genomics and post-genomics, philosophy and history of biology, and the anthropology of science.

Claire Waterton is Senior Lecturer in Environment and Society and Co-Director of the Centre for the Study for Environmental Change (CSEC) within the Sociology Department of Lancaster University.

Rebecca Ellis is Lecturer in the Lancaster Environment Centre at Lancaster University.

Brian Wynne is Professor of Science Studies at CSEC Lancaster University, and from 2002–2012 was co-PI and Associate Director of the ESRC Centre, Cesagen.

Genetics and Society

Series Editors: Ruth Chadwick, *Director of Cesagen, Cardiff University*, John Dupré, *Director of Egenis, Exeter University*, David Wield, *Director of Innogen, Edinburgh University* and Steve Yearley, *Director of the Genomics Forum, Edinburgh University*.

The books in this series, all based on original research, explore the social, economic and ethical consequences of the new genetic sciences. The series is based in the Cesagen, one of the centres forming the ESRC's Genomics Network (EGN), the largest UK investment in social-science research on the implications of these innovations. With a mix of research monographs, edited collections, textbooks and a major new handbook, the series is a valuable contribution to the social analysis of developing and emergent bio-technologies.

Series titles include:

New Genetics, New Social Formations
Peter Glasner, Paul Atkinson and Helen Greenslade

New Genetics, New Identities
Paul Atkinson, Peter Glasner and Helen Greenslade

The GM Debate
Risk, politics and public engagement
Tom Horlick-Jones, John Walls, Gene Rowe, Nick Pidgeon, Wouter Poortinga, Graham Murdock and Tim O'Riordan

Growth Cultures
Life sciences and economic development
Philip Cooke

Human Cloning in the Media
Joan Haran, Jenny Kitzinger, Maureen McNeil and Kate O'Riordan

Local Cells, Global Science
Embryonic stem cell research in India
Aditya Bharadwaj and Peter Glasner

Handbook of Genetics and Society
Paul Atkinson, Peter Glasner and Margaret Lock

The Human Genome
Chamundeeswari Kuppuswamy

Community Genetics and Genetic Alliances
Eugenics, carrier testing and networks of risk
Aviad E. Raz

Neurogenetic Diagnoses
The power of hope and the limits of today's medicine
Carole Browner and H. Mabel Preloran

Debating Human Genetics
Contemporary issues in public policy and ethics
Alexandra Plows

Genetically Modified Crops on Trial
Opening up alternative futures of Euro-agriculture
Les Levidow

Creating Conditions
The making and remaking of a genetic condition
Katie Featherstone and Paul Atkinson

Genetic Testing
Accounts of autonomy, responsiblility and blame
Michael Arribas-Allyon, Srikant Sarangi and Angus Clarke

Regulating Next Generation Agri-Food Biotechnologies
Lessons from European, North American and Asian experiences
Edited by Michael Howlett and David Laycock

Regenerating Bodies
Tissue and cell therapies in the twenty-first century
Julie Kent

Gender and Genetics
Sociology of the prenatal
Kate Reed

Risky Genes
Genetics, breast cancer and Jewish identity
Jessica Mozersky

The Gene, the Clinic, and the Family
Diagnosing dysmorphology, reviving medical dominance
Joanna Latimer

Barcoding Nature
Shifting cultures of taxonomy in an age of biodiversity loss
Claire Waterton, Rebecca Ellis and Brian Wynne

Forthcoming titles include:

Scientific, Clinical and Commercial Development of the Stem Cell
From radiobiology to regenerative medicine
Alison Kraft

Negotiating Bioethics
The governance of UNESCO's bioethics programme
Adèle Langlois

Barcoding Nature

Shifting cultures of taxonomy
in an age of biodiversity loss

**Claire Waterton, Rebecca Ellis
and Brian Wynne**

Routledge
Taylor & Francis Group

LONDON AND NEW YORK

First published 2013
by Routledge
2 Park Square, Milton Park, Abingdon, Oxfordshire OX14 4RN

Simultaneously published in the USA and Canada
by Routledge
711 Third Avenue, New York, NY 10017

First issued in paperback 2014

Routledge is an imprint of the Taylor and Francis Group, an informa business

British Library Cataloguing in Publication Data
A catalogue record for this book is available from the British Library

Library of Congress Cataloging in Publication Data
Waterton, Claire.
Barcoding nature: shifting cultures of taxonomy in an age of biodiversity loss / Claire Waterton, Rebecca Ellis and Brian Wynne.
pages cm
"Simultaneously published in the USA and Canada"–Title page verso.c
Includes bibliographical references and index.
1. Biology–Classification. 2. Biology–Classification–Social aspects. 3. DNA fingerprinting of plants. 4. DNA fingerprinting of animals 5. Bar coding. 6. Genomics–Data processing. 7. Genetic markers. 8. Biodiversity–Monitoring. 9. Biodiversity conservation. I. Ellis, Rebecca (Social scientist) II. Wynne, Brian, 1947- III. Title.
QH83.W37 2013
578.01'2–dc23
2012043533

ISBN 978-0-415-55479-4 (hbk)
ISBN 978-1-138-80785-3 (pbk)
ISBN 978-0-203-87044-0 (ebk)

Typeset in Times New Roman
by GreenGate Publishing Services, Tonbridge, Kent

Contents

Figures

Acknowledgements

The research underpinning this book was carried out between 2006 and 2009, with support from the Economic and Social Research Council (ESRC) under the responsive mode grant 'Taxonomy at a Crossroads: Science, Publics and Policy in Biodiversity' (RES – 000-23-1470) and under the programme of the ESRC Research Centre for Economic and Social Aspects of Genomics (Cesagen, 2002–2012). The authors gratefully acknowledge the ESRC's support from these sources. The research would not have been possible without the earlier forging of particular and lasting collaborations across very different domains of research and practice and we would like to thank certain of our colleagues personally: Johannes Vogel, Director of the Museum for Natural History, Berlin (formerly Keeper of Botany in the Natural History Museum, London); Gill Stevens of the Natural History Museum, London; Robin Grove-White, Director of the Centre for the Study of Environmental Change, Lancaster University; Mark Carine of the Natural History Museum, London; David Schindel, Executive Secretary, Consortium for the Barcoding of Life, Smithsonian Institution, Washington DC; Paul Hebert and Sujeevan Ratnasingham, Director and Informatics Lead, respectively, of the Canadian Centre for DNA Barcoding, Ontario Biodiversity Institute, University of Guelph. These collaborators and research partners not only gave up much valuable time to assist us in understanding the science and the politics of contemporary taxonomy and DNA barcoding. They also helped us find the right people to approach on different aspects as we carried out our study – we could not have done the research without their generous co-operation. We are also indebted to all of our interviewees, from museums, institutes and universities around the world, for speaking with us about their own research and their thoughts about contemporary shifts in taxonomic culture and biodiversity. We thank them all for their enthusiastic, patient and insightful inputs to our study. A final conference, 'Archiving and Innovating Diversity: Cybergenomics Meets The Order of Things', marking the end of the ESRC 'Taxonomy at a Crossroads' research award, was held in the Natural History Museum in London in 2009, and we would also like to thank our key speakers at that event: Staffan Müller-Wille, ESRC Genomics Network, University of Exeter; Bronwyn Parry, Queen Mary College, University of London; as well as Sujeevan Ratnasingham and David Schindel mentioned above. We are grateful to the editors of the Genomics and

Society series at Routledge, together with their peer reviewers, for giving us the opportunity to publish this ethnographic study. Sincere thanks, in particular, go to Emily Briggs at Routledge for her advice and assistance and for her calm support throughout the writing and publishing process. Three colleagues from our own disciplinary heartland in science and technology studies (STS), Kristin Asdal, Ruth McNally and Jenny Reardon, were generous enough to go over parts of the final manuscript prior to 'production' and we are extremely grateful for their comments and judgements. Of course, we authors bear full responsibility for any remaining shortcomings of this book.

Finally, we thank our respective partners and families – evidently the most important part of the entire jigsaw-in-motion – who have kept us buoyant through-out, and extended their care to embrace the completion of this writing project.

1 Introduction

The Natural History Museum in London (hereafter the NHM) curates and displays one of the most comprehensive and world-famous collections of specimens of once-living forms. Many people who walk into the museum are lured in, enchanted but also bewildered by life's variety. Some perhaps tune into a sense of reverence towards several millennia of life on the planet, only some of it coincident with human existence. Many natural history aficionados herald such museums as examples of precious 'secular cathedrals' of our time (NHM 2003). Fragile butterflies and moths carefully pinned and aligned in mahogany cabinets compete for space and attention with glass dioramic cases of perching birds and numerous other displays. The more conventional historical collections of dried, pressed, pinned and stuffed organisms spanning all five of nature's kingdoms today compete with the new Darwin Centre's spirit collection. This houses what, for some people, is perhaps the most exotic specimen of all – a giant squid, once holding a place in the human imagination as a mythical being – suspended in formol-saline within an acrylic tank, nine metres long. Contributing to the sensation conjured by the ghostly and fleshy presence of the squid is the fact that the tank was built by the same company furnishing the artist Damien Hirst with containers for his exhibits. Embodied memories and traces of lost and still-living species are ordered, preserved and classified in every corner of the vast museum. These stand for at least part of the planet's biodiversity, some of which continues to flourish in living form beyond the walls of the museum.

Less apparent to the public eye are the labyrinthine corridors, impossibly steep and twisting, narrow staircases and rows of offices, storage cabinets and laboratories behind the scenes. This is the setting for a hive of taxonomic activity. The taxonomists working at the museum – scientists who devote their activities to identifying organisms and tracing the evolutionary and bio-geographical relationships between them – draw upon the vast collections of over 60 million life-science specimens. The Natural History Museum is known, for example, amongst the world's community of cryptogrammic specialists – those individuals with an interest in lower plants such as mosses, liverworts and ferns – to house a significant collection of specimens of the delicate and endangered Killarney fern, the idiosyncrasies of which are close to the heart of one of our museum colleagues. Today the museum's specimens consist of whole organisms and of fragments of

DNA cryogenically stored at −180 °C. Each taxonomist tends to specialize in a particular group of organisms, but each knows that their specific expertise and breakthroughs in understanding contribute to the greater understanding of relations between living (and extinct) organisms on the planet.

Between 2002 and 2005 we had the privilege of working behind the scenes with a number of museum scientists. Our research at that time – a multi-sited ethnography, at the museum but also in the field with various groups – was exploring the relationships between amateur naturalists in the UK and the broader domain of biodiversity science and policy.[1] Towards the end of this research, we found ourselves, as was often the case, relaxing and chatting in the Botany common room at the top of the building. We had become accustomed to participating in conversations on broad-ranging topics, including sometimes quite heated discussions on the controversies around moss and algae identification. Such conversations were also a kind of fieldwork for us. But on this occasion we were informed of a discovery just announced which was potentially revolutionary for *taxonomy* writ large. Paul Hebert, leading a team of taxonomists, molecular biologists and bioinformaticians at the Ontario Biodiversity Institute in Guelph, Canada, was the protagonist here, and a very energetic and articulate one. He had discovered – we were told in a mixture of animated and skeptical tones – that not only could DNA be a key to understanding the similarities, differences and evolutionary relationships between species (a method established in taxonomy since the 1950s), but that a small fragment of any organism's DNA – a so-called 'micro-genome' – could be translated into a digital artefact that would look very similar to a conventional product 'barcode'. This barcode, in turn, could be banked in a global library which, when full of barcodes, could serve as a reference point for the rapid identification of every species on the planet! Paul Hebert, in interview with us in 2006, referred to this vision as *one gene = all species = all life*. The global scale of the claim and the language of renaissance, revolution, global reach, precision and speed that it promised for the science of taxonomy, sounded exciting. Hebert had introduced a palpable acceleration of rhythm and energy to the debates around this science that we had not hitherto witnessed. Taxonomy was a field with a reputation for deliberate craft-intensive attention to minute detail, especially of morphological characters, comparisons and changes. It was not known for its animation, speed and desire for dramatic centre-stage influence.

One of our colleagues at the NHM had an uncanny ability to keep his finger on the pulse of his own shifting scientific culture but also to tune into our own imaginations and referents as anthropologists and sociologists of science. He suggested to us that if we were really interested in understanding contemporary cultures of taxonomy, we should find out more about these new proposals, which he believed were set to rock the foundations of taxonomy, as he then knew it. We were inevitably enticed by this prospect, and this book is the result of a three-year (2006–2009) ethnographic study of DNA barcoding and its evolution through the development of what became known as the Barcoding of Life Initiative (BOLI).[2]

Informed both by science and technology studies (STS) and the anthropology of science, our methods and modes of analysis commit us to investigate the emerging innovation of DNA barcoding as history and culture in-the-making. The present study has interpreted this innovation as belonging to wider shifts underway in taxonomic and related scientific cultures, whilst keeping a close eye on the specific changes it is propelling in the collecting, organizing, analysing, archiving and imagined use of biological specimens and biodiversity data. We have thus explored the intimate and fluid connections between the minutiae of biological organisms (their DNA) and hopes, dreams and visions for future human and planetary existence. We observe ways in which the taxonomic community, which inherits several centuries of tussling with approaches to classification of the natural world, now finds itself caught up in the whirl of two wider intersecting domains – of the big hyper-genomic life sciences, and biodiversity science and policy.

We note here that STS/anthropology of science studies of the genomic sciences and their publics have mostly focused on the field of red genomics (the bio-medical sciences and technologies) and the therapeutic human bodily futures they conjure. Our research, on the other hand, has explored the domains of non-human biology or ecology which intersect with genomic approaches – the so called 'green genomics'. With exceptions in the area of genetically modified organisms, green genomics seems to have been largely neglected in STS/ anthropology of science. Taking many cues from STS/anthropology of science research in red genomics, therefore, we look closely at the pursuit of classifying biological organisms and we consider the relevance this may have for re-thinking past, present and future relationships between humans and the multitude of organisms with which we co-habit the planet. Our attempt to understand the science of genomics (through a 'green' rather than a 'red' lens) leads us to connect hopes for human survival and betterment with the (future) health of biodiversity and the planet as a whole. The kinds of questions we ask in this book range from those concerning the smallest details of changes in knowledge production and practice within the taxonomic community, to broader ranging questions about biology, 'life' and the human position within this more comprehensive ontological frame. What specific organic fragments, for example, do scientists trust, selected from the entire biological make-up of organisms, to communicate appropriate information and promise for the contemporary purposes of the taxonomic and biodiversity sciences, here re-rendered through genomics? What does 'life' become as it is encountered, classified, selectively represented, manipulated and mobilized within the speedy, almost frenzied, 'technopreneurial' context of contemporary genomic life-sciences? What reciprocities are crafted to maintain relationships between the innovation of DNA barcoding and long-established continuing taxonomic methodologies for ordering life? In what ways are human subjectivities – ways of being human – imagined and woven together with hopes for biodiversity protection, in a genomic barcoded future? What deep intuitions around the protection of species, but also their loss and destruction, inform the urge to archive – for universal access and use – the DNA barcodes of all species on the planet?

The research has required us to zoom in to the micro-scale of understanding the making of a digital genetic artifact – the DNA barcode – and out again, globally, as barcoders have attempted to press a single gene fragment, found in all eukaryotic organisms,[3] to reveal the diversity of all life on earth. It has involved, for example, navigating the complex geo-bio-political histories of specimen collection and archiving up to the present time, and gaining an understanding of the protocols being established to regulate the global access to and mobilization of genetic material for taxonomic research. Although sometimes overcome by the scale and heterogeneity of the sites of our new research, we also felt very lucky! It allowed us to continue working with the rich imaginative worlds of taxonomists and their organisms and we continued to frequent specimen collections around the world. It sent us to robotic sequencing laboratories and international meetings where we witnessed bioinformaticians, laboratory and database managers deep in discussion concerning the data standards required for DNA barcoding to become established taxonomic practice. It also took us to meetings of the Convention of Biological Diversity (CBD) – the powerhouse of the turbulent domain of global biodiversity policy and politics. In other words, we experienced first-hand the proliferation of intersections between taxonomic science, molecular biology, biodiversity genomics and global policy, just at the time when all of these were shifting and emerging in more intensified forms, creating a platform for new ways in which global science, and society encounters the natural world (Fischer 2005).

Paul Hebert, the lead player introduced above, provided unreserved support for our ethnography of an enterprise – DNA barcoding – that he believed would, as he put it, 'remake our relationship with life'. At the time this seemed like a grand claim coming from the traditionally quite humble domain of taxonomic science, but it whetted our appetites for more. It was with some puzzlement therefore that we soon witnessed a dampening of the energy resonating from Paul Hebert's hopeful and tantalizing promise. We had been invited to the first International Annual Barcoding Meeting held in August 2006 at the NHM, to give a short presentation on our proposed research. The atmosphere in a room packed with an international crowd of taxonomists, was a mixture of intrigue, caution and perhaps even a little contempt. What was immediately apparent to us as we grappled with difficult questions and listened to further talks, was that the taxonomic community writ-large was not going to be easily seduced by what appeared on first impression to be Paul Hebert's sound-bite approach to life. We had immediately to wake up to the fact that – perhaps inevitably – the vision which had so far captivated us would be refracted through the hopes, sensitivities and commitments of a long-established, highly diverse and fragmented global discipline.

Paul Hebert was not in fact present at this meeting but his and his team's ideas were communicated through talks provided by other taxonomists. Some individuals present wanted to know more, some wanted desperately to hold Paul Hebert to account and ask him to reduce his vision to a more realistic and modest offering to the intersecting fields of taxonomy and biodiversity science and policy. The latter group demanded an acknowledgement that DNA barcoding could not be hailed as revolutionary, but was actually a rather humble evolution of taxonomic practice in one particular direction. All that was realistically on the table, they felt, was a very simple tool which may speed up species identification and mobilize data for all to use – but this was by no means set to change our relationship with life! Crucially also, in order to make its contribution, DNA barcoding was dependent upon conventional long-established taxonomic resources (such as material voucher-specimens and existing infrastructures).

The meeting was, however, an invaluable portent for our research to come. Although slightly surprised, perhaps by our own naivety more than anything else, we were not disheartened by the potentially numbing ambivalence expressed by the taxonomists present. The sense we grasped was that if we were to step into the world of DNA barcoding, the journey would require us to navigate the eddies and flows of an entangled and emergent set of relationships between promise and tangible achievement. We had come face-to-face with one of many ambiguities that should here set the tone for what is to be explored in this book. Indeed we knew from STS and anthropology of science that it is these very sites of often unspoken tension, contradiction, ambiguity and paradox that offer rich seams for further insight – something to which Mike Fortun, more than any other STS scholar of genomics, bears witness (Fortun 2009). It is in these 'chiasmic' spaces that the richest inter-connections between past, present and future; material and virtual forms; the objective and the subjective; the universal and the particular; and the global geo- and bio-political relationships in the making of technoscience can be explored. In STS terms we must look then at the politics, oscillations and frictions that constitute the barcoding network (Barry 2001; Dugdale 1999; Tsing 2005). We need to ask how evident contradictions – for example between promise and tangible achievement – might be contained within a 'project'. We also need to consider how such contradictions and ambiguities perhaps exceed any given 'project', their substantive technoscientific contours more powerfully shaped by the wider cultural context? Furthermore, what does their navigation by scientists as public knowledge producers, thus also as material social world-producers, reveal about the current shaping and experience of so-called modernity writ large?

We hope that the memories, reflections and questions provided above offer a sense of the challenge and excitement with which we began our research into DNA barcoding and BOLI – the multi-pronged institution that has supported this emerging technoscience and its promotion in worlds of practical applications and public policy. Needless to say, much detail will follow about barcoding and BOLI in the chapters of this book. The remainder of our introduction will locate our study within sociological and anthropological approaches to the phenomenon of classification and within the history of the intersections between the taxonomic and biodiversity sciences. We will then provide chapter outlines with which readers can navigate the book.

******

Sociologists, anthropologists and historians have long been rightly fascinated by human classification practices. What holds such interest, perhaps, is the idea that classifications of things do not in themselves exhibit only a pre-existing logic in nature: rather they are of social origin and perform social roles and meanings, even while reflecting systematic attention to selected salient qualities in nature. The relationships between classes, then, are not simply to do with the correspondence between the world and the classificatory scheme but also concern the collective construction of meaning (Durkheim and Mauss 1963; Bowker and Star 1999). Such processes also deserve to be interpreted historically: whilst the making of collective meaning through classification endures as a human pursuit, it also changes in its historical detail and anchoring over time. In her essay on the history of naming the diversity of life-forms on our planet, historian of science Lorraine Daston (2004) explores precisely this point.

Daston opens her essay with a rendition of the biblical story of the Tower of Babel, from The Book of Genesis. The story reads as a warning of the ultimate human hubris in pretending that all of humankind could create, out of a context of multiple linguistic classifications, a single harmonious language. According to the story, the people of this world would build a single city-state, live together there and speak a single language comprehensible to all. Beholding this activity as an act of ultimate vanity, God put an end to it, confusing language and scattering people into their separate cultures, all over the face of the earth. The resulting chaos he named 'Babel'.

The curse of Babel reverberates powerfully in many dystopic cultural fables, including scientific ones, and perhaps most tellingly in taxonomy. The urge to discover unity in nature has been a defining epistemic and normative feature of scientific modernity; and of course, this urge has been animated by an original drive to escape religious dogma's own visions of a different unity. Daston describes the fear afflicting eighteenth- and nineteenth-century plant taxonomists, for example of the catastrophic confusion that could be wrought by the proliferation of different names for the same plant species or genus – a confusion known within taxonomy as 'synonymy'. This fear, of a bewildering lack of correspondence between names and kinds – a modern Babel – continues within contemporary taxonomy. As a retired Director of the Oxford Museum of Natural History recently expressed it, 'without a permanent single system of names, how could there ever be permanence of knowledge?'

The Director's lasting concern here reminds us of the metaphysical responsibility that taxonomists (including DNA barcoders) shoulder and of the work that has gone on, over time, to achieve collective meaning. Daston describes precisely how taxonomists actually manage the tension between, on the one hand, keeping names for organisms stable (and therefore able to be collectively known and understood) and, on the other hand, allowing for change in naming and classification, as understanding of species, their boundaries and their inter-relations developed. This tension has to be actively managed through collectively agreed conventions. The 'type specimen' is one such convention, designed to maintain a stable, collectively understood correspondence between a name and a natural kind, and as such an essential part of taxonomic practice with significant symbolic value. Without such conventions, what Daston calls the 'St Vitus' dance of scientific change' – the inevitable theoretical and technological shifts in the bases of classification – would corrode delicately transmitted common understandings, and Babel would return.

In what follows in this book we see BOLI as a contemporary taxonomic initiative shouldering complex ongoing histories of shifting taxonomic practice and commitment, all of them replete with intricate conventions enabling collective work. Such taxonomic histories are elaborate and complex and we do not enter into their details here. We note, however, that so-called natural order(s) have historically been perceived and generated through quite different lenses of enquiry, reflecting different combinations of purpose, and through different conventions, with implications for collective meaning. This historic reality has contributed to a certain epistemic turbulence which continues today as a characteristic feature of taxonomic cultures (Foucault 1966; Dean 1979; Hull 1988; Stemerding 1991; Ridley 1986).

Scholars reveal, for example, that the portrayal of natural order(s) by taxonomists depends very much upon what properties of organisms are considered to be indicative of order (Hesse 1974). As Foucault's (1966) oft cited work on the history of natural historical pursuits demonstrates, the classification system developed in the seventeenth and eighteenth centuries known as the Linnaean hierarchy[4] essentially operated – and continues to do so today – as a linguistic device equipped to lay out a grid of the observable (morphological) physical qualities of organisms, organized according to the measurements of similarities and differences between them. It is this visually perceived ordering of similarity and difference that Foucault sees as exemplary of the 'classical episteme' of the time. An organism's place in the world became defined in that it 'exists in itself only in so far as it is bounded by what is distinguishable from it' (Foucault 1966: 158).

From the beginning of the nineteenth century, a growing sense of dissatisfaction with the limits of Linnaean linguistic naming and classificatory tools encouraged natural historians to explore the temporal complexities of the relationships between organisms, and they began to present natural orders as evolving states, thus with a propensity for further change. After Darwin's 1859 evolutionary theory became established in the late nineteenth century, an interest in inheritance and common lines of reproductive descent became salient, and complemented the visual detection and analysis of morphological characters in the labour of classificatory organization. As Foucault notes, these shifts, together with Cuvier's focus upon the vital processes or functions of parts of organisms as integral to their classification, propelled the life sciences to develop from their practices of naming and differentiation, towards temporal dimensions of descent and inheritance, and functional qualities which might influence evolutionary 'fitness'. From the nineteenth century onwards, whereas natural historians and taxonomists sought to uncover the large-scale patterns of living nature, through collecting in the field and classifying in the museum, 'modern' biologists in their laboratories sought to penetrate the internal workings of the living organism to discover their fundamental causes and functions. Gradually, the emerging disciplines such as biology, zoology, ecology and genetics came to dismiss the quaintly outdated methods of taxonomy as 'unscientific' – involving, so it was perceived, mere description, naming and 'book-keeping'. Taxonomy was seen as neither hypothesis-testing nor experimental – the two key epistemic characteristics rapidly becoming established as the hallmarks of professionalized 'proper' science, such as biology.

Twentieth-century taxonomy inherited the consequences of these epistemic rifts and, according to many accounts, became increasingly fragmented and introverted as a discipline. Technological innovation, without being determining, further complicates this evolving narrative. From the mid-century onwards, a continuing dependency on the visible morphological characteristics of specimens was complemented by the use of DNA sequences as a new form of classificatory currency (Avise 1994; Hull 1988; Ridley 1986). DNA taxonomy, as the field became known, introduced powerful molecular and computing methods for indicating common ancestry and species histories and was thus hailed as harbouring revolutionary potential in terms of its ability to reconfigure species boundaries and bring a more sophisticated and higher resolution approach to understanding the descent of species.

By the end of the twentieth century, taxonomy, like most areas of the life sciences, was being reframed by state-of-the-art genomic approaches, increasingly relying on DNA and molecular techniques and powerful computer technologies to 'split and lump' the natural world. These methods of analysis built on the 1990s 'genome revolution' assumption that molecular genetic sequences were the code of life, deterministic of all which was thought to follow (Keller 2000; Kay 1993; McNally and Glasner 2007; Oyama 2000). This genomic reframing enabled an acceleration of taxonomic practice and allowed the discipline to expand its scale and speed of inquiry and delivery. The epistemic consequences of this shift were not altogether transformative for taxonomy however. They simply built upon disputes internal to a community split by allegiances to different schools of classification and their concomitant philosophies (Hull 1988; Ridley 1986). Scientific accounts of the relationship between the world and the name had always been and continued to be fragile and contested, in a way which the genomics revolution did not seem able to overcome.

Taxonomy entered the twenty-first century wringing its hands in self-reflective concern at the fragmentation and lack of standing (including funding, and new recruitment) of the field. These domestic disciplinary anxieties resonated from the early 1990s onwards with an independent set of concerns coming from science policy, and public policy beyond this, over global biodiversity loss and how to harness science including taxonomy, to respond. Taxonomy was deemed to be unfit for purpose, in face of these new expectations. But before we go on to describe the 'crisis' within the discipline that many felt to be chronic and in desperate need of repair, we make a further short point relating to taxonomic history. BOLI inherits not only epistemic conventions, but also geo- and bio-political histories. In this book we explore a range of ways in which BOLI, often unwittingly, builds upon controversial legacies through its material, digital and socio-political efforts to collect, analyse and archive global life-forms. An obvious example of bio-political configurations in the taxonomic context is presented by (neo)colonial practices of collecting and mobilizing natural specimens, often in the form of genetic material (Hayden 2003a; Parry 2004). BOLI requires an unprecedented diversity and quantity of biological organisms in its efforts to (re-)archive all life forms as (simultaneously) material specimen, DNA sequence, digital information and corresponding barcode. Such organisms derive from nature in 'the field', as well as from existing collections in museums and laboratories across the world. As such, BOLI builds upon centuries of taxonomy as a globally distributed scientific field of practice, collecting and archiving a highly varied quality of material samples, sources and original environments across life and past life; but how it reconstructs such histories, and how this reflects and reinforces its own cultural practices and relations, has been an important aspect of our research. This also has important bearing on how we reflect more broadly upon our findings, in our final chapter.

In this book we ask how BOLI, as a partial 'genomicization' of the taxonomic discipline, has intensified and scaled up the distribution of practices, between field, lab and database? How has DNA barcoding introduced a further concentration to practices of collecting and curation inherited from earlier centuries of taxonomic pursuits? How has BOLI produced new and contemporary regimes of labour and distribution? How has the celebration of the apparent power of DNA – in this case a segment of one gene, so tiny a fragment of a genome – to determine, and represent 'life' in the large, been invested with practical meaning and promised public effects, by BOLI?

Through its quite urgent need to negotiate access to material nature, BOLI has built and continues to develop, for example, collaborations with local communities in bio-diverse areas of the world. This has meant that it has inevitably had to pay particularly careful attention to the global biodiversity policy framework – the 1992 UN Convention on Biological Diversity and its Access and Benefits Sharing Protocols – in order to navigate a carefully regulated conduit to this material and its human owners and stewards. BOLI has not only relied on 'fresh' biological material however. It has also had to gain access to a range of past life as 'dry' or frozen specimens collected, ordered and cared for in natural history museums and herbaria. This has also required that the initiative establish new forms of collaboration with museum managers and curators. BOLI's practices in these regards layer tightly upon colonial and neocolonial histories of access, collecting and accumulation in the pursuit of a complete natural archive of biodiversity, often tinged by association with contested processes of commercial exploitation.

Crisis! ... In both biodiversity and taxonomy

A more immediate context for the emergence of BOLI becomes important as we introduce its avowed aim to create a complete archive encompassing all of life's diversity. As we have suggested, around the beginning of the twenty-first century there existed considerable upheaval and 'crisis' within the taxonomic sciences. But this was also true for the wider and more disparate biodiversity sciences and their policy arenas. Mounting concerns over 'The Sixth Great Extinction' of global biodiversity (Pimm and Raven 2005), together with the naming of a distinct new geological epoch referring to the comprehensive material domination of nature's fate by human activities ('The Anthropocene' (Crutzen and Stoermer 2000)), served as a potent backdrop to biodiversity experts' assessments of the rates of global biodiversity loss (Royal Society 2003). Scientists began to discuss with increasing angst their lack of knowledge about the number of species inhabiting the global biosphere: estimates were of the order that, of the 10 to 100 million species thought to exist, only about 1.5 million had by then been recorded and identified.

This degree of ignorance of global biodiversity was seen to be accentuated by two further realities, seen as the taxonomic and biodiversity crises respectively: first, the snail's pace of science's painstaking accretion of species knowledge, to the 250-year-old, slow, labour-intensive and skills-intensive discipline of taxonomy; second, the accelerating rate of loss of the very biodiversity of which we are largely ignorant, but which it is recognized may be vital for ecological and human sustainability. As if all this unmapped and neglected extinction of global life were not alarming enough, scientists involved in biodiversity research, going far beyond taxonomy alone, were embarrassed by the stark realization that for all the well-funded scientific efforts over the last two decades to know global biodiversity better, and in more accessible forms, the impressive increase which had occurred in scientific and policy knowledge had enjoyed little-or-no beneficial protective effect over biodiversity-destruction. The common belief that more science equals better policy was manifestly failing. As documented discussions during the so-called International Year of Biodiversity (2010) at the Convention of Biological Diversity attest, efforts at mapping and protecting nature have been virtually impotent in curbing rates of biodiversity loss. The mantra 'we need to know all species in order to save biodiversity' was looking increasingly incapable of living up to its promise, and in several reinforcing respects.

Thus, for many of those involved, the taxonomic and biodiversity crises, entangled in many ways, meant that taxonomic change, even renewal, was almost essential for the self-respect and continuation of the discipline. The technoscience that BOLI promotes, DNA barcoding, was advanced from 2003 onwards as a way to radically speed up and even 'democratize' the identification–classification work of conventional taxonomy. One of its many premises, the value of which we already question above, was that the urgent task of identifying that vast reservoir of unknown species before they are extinguished forever, would itself be a protective act. As Meier reflected in 2008, it was hardly surprising that an initiative such as the DNA BOLI was founded at such a time:

> Traditional taxonomy is in crisis. The literature is often inaccessible, electronic data management is in its infancy, retiring taxonomists are leaving numerous 'orphan' taxa behind, few students are entering the field, fewer yet are being hired, and academic administrations focusing on impact factors are abandoning the field altogether. Yet taxonomists are more urgently needed than ever, given that 90 per cent of all species remain undescribed, environmental impact assessments require accurate taxonomic information, and conservation managers critically need precise species data. However at

the current rate traditional taxonomy will require more than 940 years before all species will be described, because the doubling rate for species descriptions is quickly slowing. Under these circumstances, it does not come as a surprise that biologists are looking for new techniques that can speed up the process of describing and identifying specimens. It is also not surprising that these biologists are looking at molecular techniques because they have successfully revitalized and significantly improved our understanding in many fields of biology.

(Meier 2008: 96)

During our research we have often pondered about the purchase that fears of biodiversity loss hold over scientific and public communities. Questions we have found it important to ask here include: What does this very specific anxiety reveal to us about the fears and prevalent myths of this age? What is the assumed role of science in shouldering responsibility and shaping far-reaching solutions? How are questions of (ways of) knowing, of caring and of action inter-related?

Promises

In 2003, BOLI promised somewhat ambitiously to spark a renaissance for the taxonomic and biodiversity sciences combined, by 'reinventing' taxonomy as a precise and accurate high-throughput molecular science. The kind of taxonomy advocated, DNA barcoding, would become a publically accessible, user-oriented, policy-responsive and globally standardized, IT-intensive science. To its proponents the future effectiveness of this science was almost a given. Such ambitions seemed credible with the announcement of the 'discovery' of a fast, reliable, automatable and cheap process for the genetic identification of species (Hebert *et al.* 2003a and b). But DNA barcoding was different from DNA taxonomy, as we explore in Chapter 2. The so-called DNA barcode was a sequenced short stretch of DNA universally present in all (eukaryotic) species (cytochrome oxidase 1 – CO1, consisting of 648 base pairs) which, it was claimed, would work faithfully through its imagined uptake as a global standard to identify species-specific sequence variation. Although this technique was initially trialed only for animal species, by 2005 it was thought that DNA barcoding would deliver a universal toolkit and web-accessible framework for rapid molecular diagnostics for all species. What was being proposed was a universalized form of 'horizontal genomics'. The DNA barcode thus had the practical and rhetorical force associated with much of genomics; the idea that one gene = all species = all life worked to ignite the imagination of many taxonomists and perhaps more importantly, their funders and user-communities.

BOLI's agenda was multiple and ambitious. It was clear from the outset that DNA barcoding was promising to do more than identify and name natural life's myriad species much faster and more cheaply, thus rescuing taxonomy from its sense of futility and crisis. Indeed, in its scope and ambition, barcoding seemed to accompany an array of modern technosciences claiming to tackle the many grand challenges confronting human society in the opening years of the twenty-first century, such as climate change and its human causes, global food security and sovereignty, democratic freedoms and responsibilities, and the synthesis of life itself. But this ambitious promissory agenda was explicitly not only technical-scientific (while being substantially experimental), but human too. DNA barcoding, as an integrative state-of-the-art DNA sequencing and informatics-based genomic technoscience, promised not only faster, cheaper and more citizen-inclusive global identification, classification and recording of species, but also – and thanks to its particular technoscientific form – a global moral transformation of humankind and its evident indifference to nature and biodiversity's destruction, through a revitalized access to and knowledge of nature. It is important to note here that E.O. Wilson's notion of *biophilia* (1984) – conveying an almost millenarian sense of a need to 'return' to a primordial human connection with the natural world – has had a profound influence upon the hopes and rhetoric of BOLI. It has been translated through a concept of 'bio-literacy' (Janzen 2004b, 2005), which promotes the idea that true love and connection with nature is best enabled through an ability to differentiate between, and hence name and identify, natural species. Barcoders' responsibilities thus extend beyond those associated with taxonomic classification, to more socio-political and even cosmic processes (Helmreich 2009; Toulmin 1982).

******

At the beginning of this introduction we noted the analytical value for STS and anthropology, of tuning into domains of scientific practice and imagination as sites of tension, ambivalence, contradiction and occasional paradox. The initial contradiction we highlighted – and several more will be emphasized in chapters to come – was the growing sense gained by us as ethnographers, but also by the taxonomic community, that a wide gap seemed to exist between the profligate and quite rousing claims of the first announcements of DNA barcoders, and the rather more prosaic realities of their achievements and of those of biodiversity science more broadly. It is also relevant that the barcoding practitioners themselves are quite aware of these troubling contradictions. Before introducing readers to the chapter outlines, we reiterate here that our approach is to reflect on these contradictions inherent in and emerging from the intersections between the taxonomic and biodiversity sciences. There are places in the book where we are more overtly questioning and critical of BOLI's practices and imaginaries than we are in others. However, STS and anthropological analysis also encourages us to see these ostensibly contradictory qualities as mutually dependent properties of the kind of technopreneurial enterprises making their way under prevailing political economic and cultural conditions (Franklin 2005; Fortun 2009; Taussig 2006). This raises questions as to how to regard these surprising qualities of our technoscientific field. Are the apparent excesses of transformational promise for example, properly matters of bare criticism from social scientists like ourselves, implying that they could be directly overcome and superceded by their practitioners, if they were to acknowledge them? Or are they, perhaps, more deeply constitutional of this technoscience, as it is emerging and being formed (and reformed) in this particular historical and cultural context? Without at all denying the responsibility of such technoscientific practitioners, in adopting more of the latter approach than the former we do also suggest in the final chapter, a different and perhaps more adventurous agenda of responsibility, and opportunity, for DNA barcoding.

We are aware as we finish writing this book that it has been moments of disconcertment (Verran 2001) we have experienced during our fieldwork, and in subsequent analysis and writing up, which have alerted us to some of our own assumptions about the inner and outer workings of BOLI, and what they tell about science-in-the-making, and thus world-building, as contemporary social and cultural practice. We note here, as a matter that touches on the responsibility and accountability of the ethnographer, that our study and our interpretation of barcoding may, or may not, resonate with those in the scientific and taxonomic communities who have so very generously helped us with our research. We do not expect our STS- and anthropology of science-inspired intellectual framing of the study, for which we are ultimately accountable, to map neatly onto that of those involved in the innovation of barcoding itself. Nevertheless, we hope that this text, coming from its own disciplinary tradition that asks critical questions around the deliberate and the inadvertent normativities of science, is also a way of continuing – and yes, structuring, in some respects – the conversation we have enjoyed to date with our partners in the taxonomic and biodiversity sciences and beyond.

Introduction to the chapters

Between Chapters 1 and 2 we have inserted a brief 'Technical interlude'. This is a guide to introduce the reader to the stages whereby a biological organism becomes a DNA barcode banked in the Barcoding of Life Database System (BOLD). The Technical interlude can also be read as an insight into the way that life-forms can be materially and semiotically detached from their previously bounded bodies (Helmreich 2009: 280). We use Helmreich's concept of 'biology unbound' later in the book. Such detachment, of course, involves extensive reattachments; reorganizations and reconstitutions as those organisms become standardized digitalized DNA barcode representations within a singular global format.

Chapters 2, 3 and 4 all work to construct a narrative about how DNA barcoding emerged and continues to develop as a product of the taxonomic and biodiversity crises, from 2003 onwards. They explore, in different ways, what has been required for the proposals of barcoding to become 'do-able' science (Fujimura 1987). Chapter 2 describes in some detail the scientific proposals that accompanied the introduction of barcoding as a new form of taxonomy in the early 2000s. Competing visions and arguments for and against barcoding highlight the sensitivity of the overarching discipline of taxonomy to this new, universalizing, micro-genomic system. The chapter describes how, in order for DNA barcoders' extraordinary visions to become ordinary practice, certain subtle devices – forgetting, conciliation, naturalization and 'infrastructural inversion' – were necessary.

Chapter 3 is the first time that we look in detail at CO1, the segment of mitochondrial DNA which was initially loaded with the responsibility to differentiate all animal life. The chapter examines the construction of this marker as a ventriloquist for nature (Haraway 1997) and draws attention to the intricate crafting required for its naturalization as the key to faster, cheaper, more efficient identification of species. The second half of the chapter considers BOLI's efforts to accommodate certain 'speed bumps' that have to be navigated by barcoders as specimens from the plant kingdom come to take their place alongside the animals that complied so effortlessly with CO1. Frictions (Tsing 2005) in BOLI's standardizing vision both slow down and enrich the barcoding enterprise.

Chapter 4 introduces the reader to the flow of biological materials required by BOLI to provide proof of concept and to fill the Barcoding of Life Data System (BOLD). Here we focus attention on the historical legacy of natural history collecting upon which BOLI depends, and builds, as it faces the need to access materials from a range of different locations, institutions and actors across the globe. The chapter explores the labour and politics involved in BOLI's procurement of living and dead matter and looks carefully at the ways that BOLI works to extract a new nature, from a dense variety of natural–cultural entanglements.

Having laid out the basic story of barcoding's emergence in Chapters 2, 3 and 4, Chapters 5, 6 and 7 begin to work more analytically to explore the social, cultural and philosophical dynamics of DNA barcoding. Chapter 5 interrogates the ambiguities and frictions inherent in efforts by BOLI to extend barcoding practices and networks across the globe. BOLI is analysed as a fast-paced, future-oriented technopreneurial enterprise (Ong 2005) that, despite its biodiversity protection oriented mission, is entangled in contemporary fusions of nature, culture and capital.

Chapter 6 uses a philosophical lens to examine the archiving of DNA barcodes in BOLD. The chapter explores the drive to create an entire corpus of humanity's knowledge of life through BOLD, and connects this drive to issues of care for, and loss of, biodiversity. It considers the way that the BOLD archive holds together particular ideas about future natural–social orderings and facilitates new collaborations around the 'unbound' 'biosemiotic' objects in its care.

Chapter 7 turns to look at BOLI as a key element of a hopeful and expansive cosmology that links knowledge of species to concepts of human bioliteracy and thence *biophilia*, or love of nature (Wilson 1984). It explores the ways in which BOLI, like many other genomics enterprises, harbours salvationary, redemptive aspirations; and we examine how tensions between the mundane uses of barcoding and its salvationary potential are upheld through systems of signage and promising, as embodied in the very form of the barcode itself.

Chapter 8 considers DNA barcoding's claims to utility, looking more closely at the mundane uses of barcoding in the context of a 'pluriverse' (Latour 2010), a world in which taxonomy has extended its networks to engage in the politics of biodiversity protection. The chapter explores barcoding's role as a speedy, efficient genomic 'service' for the identification of unknown organisms or parts of organisms in areas of regulation, public health and environmental monitoring. It considers, on the one hand, how this production of knowledge, as a service, requires a subtle understanding of the uncertainties of this technoscience as well as its reciprocal ties to taxonomy. On the other hand, the chapter also bears witness to the fact that DNA barcoding is developing in ways which extend beyond traditional taxonomic concerns into an imagined future where hybrid, multiple environmental samples (such as the mud from lake bottoms) will be analysed in DNA sequence terms through index numbers rather than species names. Barcoding's utility is thus in some ways steering it away from taxonomy.

We return in Chapter 9 to the several overlapping processes of apparent contradiction that inhere in BOLI's endeavor to barcode and archive all life-forms on the planet, and redeem society's destructive alienation from nature. We also raise some open-ended questions, to which we do not necessarily have answers: how should we make sense of the contradictions and ambiguities which seem to permeate the very life of BOLI? Should these qualities, which we argue throughout are not ephemeral to BOLI but constitutive of it, and which shape its epistemic culture, be treated as politics, or poetics, or perhaps both, in the wider, public frame of modern technoscience? Some of the unspoken tensions, excesses and contradictions which we have identified as essential to BOLI, and DNA barcoding, are known to practitioners, but remain unexplored by them. Modern technosciences like DNA barcoding strive for authority partly through discursive purification of their own complex entanglements of nature and culture, while continually extending and deepening those entangled hybrid forms. Drawing upon anthropological readings of neocolonial power and of technoscience, we demonstrate how the historical context of technoscience is selectively engaged in its formation. By this we mean that exaggeration of scientific promise for an increasingly insecure and perhaps overly receptive global society, hype which has been especially intense around genomics, and roundly criticised by many (Nightingale and Martin 2004; Rose and Rose 2012; Brown 2003; Fortun 2001; Sunder Rajan 2006), is a part of the political-economic and cultural context of genomics, and not only for commercially interested versions of this big science field, but even for its protection-oriented modes, such as BOLI set out to be. Such exaggerated technoscientific belief, ambition and public promise has been a striking feature of big biology in a neoliberal and neocolonial age, and unsurprisingly, this has shaped its own development. In the final chapters of this work we explore how this under-spoken ethos of human *moral* as well as instrumental promise in the genomic technoscience of barcoding nature could be rendered more modest, and thus perhaps more robust and effective, by reading its claims, as a public matter, in a different register. This would, of course, involve transformations in what we experience as modern technoscience.

Technical interlude

We include here a brief technical guide to the DNA barcoding of life. We have adapted the text below from introductions to DNA barcoding on CBOL, BOLI and iBOL websites,[1] giving as much detail as we think will be useful to the reader. This Technical interlude may also serve as a useful reference point later on, for readers who want to go back to the basic premises, processes and translations involved in allowing a biological organism to be also a red, blue, black and green coloured barcode. We hope that the descriptions below help the reader understand how it is that organisms become 'unbound' (Helmreich 2009) through DNA sequencing and barcoding, or to put it another way, how it is that they become simultaneously organism and barcode-sign, the 'material-semiotic' objects (Haraway 1997) that we refer to throughout this book. The technical procedures are of course only part of the story – for a more synthetic account we refer the reader to the remaining chapters!

DNA barcoding

DNA barcoding was proposed in 2003 by Paul Hebert and colleagues at the University of Guelph, Ontario, Canada, as a new system of species identification using a short 648 base-pair (thus rapidly-sequenced) section of DNA from a standardized region of the genome. The main concept of DNA barcoding is that such a small piece of an organism's DNA can be isolated, sequenced and matched to generate an identification of that species, usually to a previously classified distinct Linnaean species name. Initially, DNA barcoding was trialed on animal groups that had already been identified in taxonomic science. From 2005 onwards, the system has been extended to encompass plants, bacteria, fungi, protists and viruses. DNA barcoding is predominantly a system that aims to speed up the identification of species already known to science, rather than a system that sets out to discover or delineate species that are unknown to science.

Animals

DNA barcoding was proposed as a global taxonomic system when a 648 base-pair region in the mitochondrial cytochrome c oxidase 1 gene (termed CO1 in scientific publications, and referred to as CO1 hereafter in this book, see Figure T1.1a

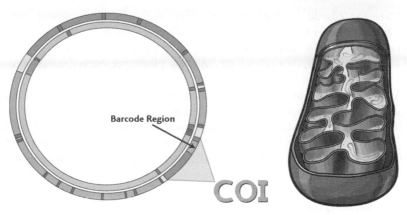

TI.1a and TI.1b Image of CO1

Source: Created by Suz Bateson for the Ontario Biodiversity Institute, University of Guelph. Permission to reproduce image granted by Suz Bateson.

and b) was found to be useful in quickly, cheaply and reliably allocating specimens to species identity.

CO1 is present in the genome of all animals and its analysis was found in 2003 to be effective in differentiating birds, butterflies, fish, flies and many other animal groups to a species level identification. The advantage of using CO1 is that it meets three crucial requirements for barcoding. First, it is a short enough gene segment to be sequenced quickly and cheaply. Second, it is long enough to identify variations between species. Third, using CO1 analysis, the variation that can be detected between different species is routinely found to be more marked than the variation within the same species. Therefore barcoding, based upon CO1 analysis, establishes a clear signal of species difference and does not cloud species identification with signals of intra-species difference.

Plants and other organisms

It was realized in the early 2000s that the CO1 barcode is not effective for identifying plants or fungi. The genomes of bacteria, viruses and protists have also posed a number of complications for the use of the CO1 barcode and taxonomists and DNA barcoders continue to discuss whether CO1 or other gene regions should be used for barcoding these organisms. In the case of plants, CO1 does not work because plants evolve more slowly than animals. This means that the mitochondrial genome has low rates of nucleotide substitution. Also, the mitochondrial genome has a rapidly changing gene content and structure that does not preserve changes to its make-up over time. The combination of these factors means that CO1 cannot signal similarities and differences between closely related plant species. However, a search for genes that might yield precise species identification from 2005 onwards brought forth several possibilities in a different region of the genome – the 'plastid' genome (or 'chloroplast' region as it is commonly called). After nine candidate regions were tested and their attributes debated, two regions were selected in 2008: 'matK' and 'rbcL'. When these two regions are sequenced and analysed together, they can reliably allocate plant specimens to an established species identity. matK and rcbL have therefore been approved as the standard DNA barcoding regions for land plants. Like CO1, they can be used to rapidly and cheaply match specimen DNA to an already documented species identity. In the case of fungi, the most common marker used for years before DNA barcoding is known as the large ribosomal DNA 'Internal Transcriber Spacer' (ITS). Although this marker does not work for all fungal groups, it is now recognized by the Barcode of Life Data Systems database (BOLD) (the DNA barcode database – see next section) as the default identification tool for fungal barcodes. Whereas BOLD's infrastructure was initially designed to process and analyse only CO1, it can now process the multiple genes (BOLD version 3.0) indicated above.

How does it work? From organism to barcode

Species identification using DNA barcodes starts with the animal, plant or other (e.g. fungal) specimen. Barcoding projects obtain specimens from a variety of sources. Some are collected in 'the field'; others come from existing collections in natural history museums, zoos, botanical gardens, herbaria and seed banks.

In the laboratory, technicians use a tiny piece of tissue from the specimen to extract its DNA. The barcode region (CO1 in animals, or matK and rbcL in plants, ITS in fungi) is isolated from this DNA. It is then replicated using a process called PCR amplification, the standard method for amplifying insufficient quantities of specimen DNA for sequencing. Lastly it is sequenced using automated sequencing robots and agreed-upon standard chemicals and procedures (timings, etc.). The DNA sequence is represented by a series of letters C, A, T and G. These letters correspond precisely to the ordering of nucleic acids in the specimine's DNA – cytosine, adenine, thymine and guanine – in the sequence under consideration.

The example given on the iBOL website, and adapted here, is that of a bird, the Arctic warbler. If you were to write down a CO1-generated sequence of an Arctic warbler (a bird that has been classified and named by taxonomists as *Phylloscopus borealis*), it would look like this: CCTATACCTAATCTTCGGAGCATGAGCG GGCATGGTAGGC … its corresponding barcode image is normally displayed in colour (represented in Figure TI.2 in different shades of grey), where C is blue, T is red, A is green and G is black.

Figure TI.2 Image of Arctic warbler DNA barcode

Source: Reproduced from http://ibol.org. Permission to reproduce image granted by Sujeevan Ratnasingham, Canadian Center for DNA Barcoding, Ontario Biodiversity Institute, University of Guelph.

If you then went on to generate a CO1 based sequence of another species of the warbler family, say a willow warbler, the sequence would be different and the corresponding coloured barcode would appear differently. Computers can be used to detect this difference automatically, in both sequences and barcodes. They can also analyse and represent difference in various ways using different algorithms such as distance analysis, neighbour-joining analysis and so on. Experienced barcoders can recognize the difference between some barcodes and estimate which species they refer to byd eye alone! Differences in DNA barcodes can be translated into what taxonomists call 'trees', showing genetic distances between species, as Figure TI.3 indicates.

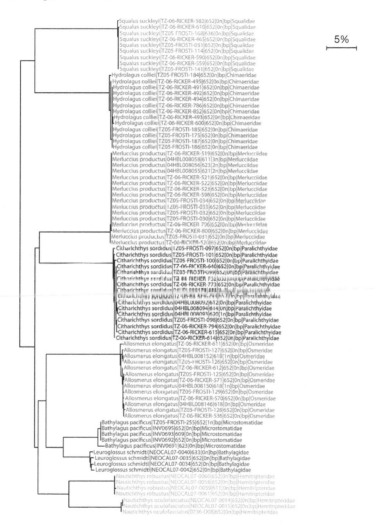

Figure TI.3 Neighbour joining tree of the Pacific fishes of Canada

Source: Permission to reproduce image granted by Sujeevan Ratnasingham, Canadian Center for DNA Barcoding, Ontario Biodiversity Institute, University of Guelph.

Once a barcode sequence for a given species has been obtained, it is archived in BOLD. BOLD is gathering such sequences as a reference library of DNA barcodes which correspond to known species names. As well as a reference library, BOLD is also a 'workbench' where analysis of difference can take place (see Chapter 6).

The DNA barcoding database: BOLD

BOLD is a searchable repository for barcode records. It is at the centre of what barcoders call the barcoding analytical chain, or 'pipeline', and it stores specimen data and images as well as sequences and trace files. BOLD can also analyse incoming data. It provides an identification engine based on the current barcode library and monitors the number of barcode sequence records and species coverage.

Barcoding institutions

BOLI. The Barcode of Life Initiative began in 2003 with Paul Hebert's announcement that species could be told apart by using a very short gene sequence from a standardized position in the genome. http://www.barcodeoflife.org.
   Core partners within BOLI are:

- **CBOL**. The Consortium for the Barcode of Life is an international initiative devoted to developing DNA barcoding as a global standard for the identification of biological species. It was initiated in 2004 and directed from the Smithsonian Museum, Washington, USA. http://www.barcodeoflife.org/content/about/what-cbol.

- **iBOL**. The International Barcode of Life project is the largest global consortium of DNA barcoding projects, labs and networks to date. It was initiated from the University of Guelph, Canada, in 2010. http://ibol.org.

- **BOLD**. The Barcode of Life Data Systems is an online workbench that aids collection, management, analysis and use of DNA barcodes. It consists of three components (MAS, IDS and ECS) addressing the needs of various groups in the barcoding community. http://www.boldsystems.org.

- **GenBank®**. Genbank is a genetic sequence database, run through the USA's National Center for Biotechnology Information (NCBI) and is part of the International Nucleotide Sequence Database Collaboration, which comprises the DNA DataBank of Japan (DDBJ), the European Molecular Biology Laboratory (EMBL) and GenBank at NCBI. It consists of an annotated collection of all publicly available DNA sequences, including those in BOLD. http://www.ncbi.nlm.nih.gov/genbank.

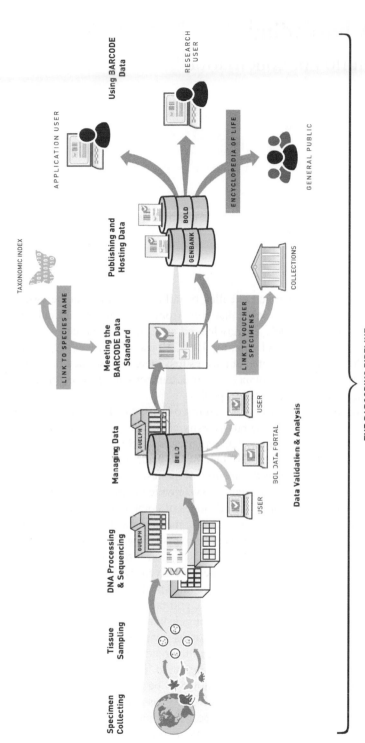

Figure TI.4 BOLD within the barcoding pipeline

Source: Figure reproduced by permission of David Schindel of the Consortium for the Barcode of Life (CBOL).

2 DNA barcoding

Revolution or conciliation?

Biodiversity is in crisis, and taxonomy is now in vogue again. This newfound
enthusiasm may well lead to upheavals in the nature of taxonomy itself.
(Mallet and Willmott 2003: 57)

In our introductory chapter we noted a feeling of failure infecting both the biodiversity and taxonomic sciences, a sense that grew from early concerns articulated
in the 1980s to a situation widely acknowledged as a crisis by the early 2000s
(Wilson 1985, 1992; Wheeler and Cracraft 1996; House of Lords 1992, 2002;
Royal Society 2003). As official rates of species loss increased, so inevitably did
the pressure felt by taxonomists themselves to document nature faster than the
rate of its ongoing extinction, species by species.

The roots of the taxonomic crisis were felt to go beyond its failure to meet the
perceived knowledge demands of the biodiversity crisis, however. From the mid-
twentieth century onwards, taxonomy had experienced increasing fragmentation
as a discipline (Hull 1988; Ridley 1986). One of the main components of this, as
many taxonomists were aware, was the upsurge of interest in the common ancestry
of species made popular by the phylogenetic systematic theories of Willi Hennig
(1966). These were enabled through the advances of computer systems from the
1960s onwards and perceived to be a rigorous, mathematically based, hypothesis-testing science. One effect of this was to overshadow the more traditional
taxonomic tasks of species delineation and description through (morphological
and DNA) character analysis. Taxonomy became riven by different principles
of character selection, which had begotten, broadly speaking, three main schools
of classification (the evolutionary, the phenetic and the cladistic schools (Ridley
1986: 163)). Disputes thrived in the ensuing struggle between different taxonomic
schools, as each one claimed the objectivity of its own philosophy of character
selection (Ridley 1986: 2). Thus, even though the broad category of 'descriptive
taxonomy' as Wheeler (2004) calls it (in distinction to 'phylogenetic taxonomy',
more interested in species ancestry), was known to be the necessary underpinning for understanding biodiversity and biodiversity loss, it came under repeated
attack from the 1990s onwards as too slow, inefficient and subjective to create an
effective evidence-base for biological diversity with global concerns mounting

over rapidly accelerating rates of biodiversity loss. The ironies of this situation in the so-called 'information' or 'digital' age, a time when the production of human knowledge was accelerating at an exponential rate, only served to increase the powerful sense of stagnation and crisis in taxonomy:

> On average the sum of human knowledge (estimated by titles of books published) doubled every 33 years before the information age. The time was reduced to c. 6 years in the 1970s, 1.5 years in 2000, and by some estimates will soon double several times per year. In sharp contrast, the doubling of taxonomic knowledge has not kept pace. Before the Enlightenment our knowledge of species doubled about once every 400 years. By the late eighteenth century, in the time of Linnaeus, our taxonomic knowledge doubled every 50 years. By the mid-twentieth century, this had slowed to doubling every 100 years and now is closer to 200 years. This should be an unacceptable trend in the midst of the biodiversity crisis.
>
> (Wheeler 2004: 572)

The feeling of demoralization in taxonomy was sharply accentuated for some by the dynamism of investment, data production and extravagant promise of social benefits coming from the genomic biosciences. The 'crossroads' for taxonomy looked clear for some: one way was to continue the same trajectory via a fragmented global taxonomic effort, hampered by schisms such as that between phylogenetic and more descriptive taxonomy highlighted above. This would mean painfully slow progress in new species delineation, and most likely further a decline in funding and recruitment. The other option was to change direction radically, embrace the multiple emergent technical and social possibilities associated with genomics and bioinformatics, and inject new life into the discipline. At the same time, this would probably mean gains in funding, new technologies and facilities for collecting, storing and archiving life's diversity. A 'revolution' was the way that several taxonomists described what was needed.

Articles addressing taxonomy's future proliferated in the early 2000s as part of a fervent context emanating from genomics and the biosciences which was highly nourishing of new visions, as well as for debate and contestation.[1] This chapter looks in some depth at three quite extraordinary proposals for taxonomy emerging in the early 2000s, each in their own way charting the prospect for a new 'taxonomic system' to revive the methods as well as the fortunes of taxonomy, and to establish it as a bonafide 'twenty-first-century science'.

We have selected these proposals from the midst of a flood of debate and critique appearing in journals such as *Nature, The Philosophical Transactions of the Royal Society of London, Systematic Biology* and *Trends in Ecological Evolution* because of their extraordinary ambition and because of the sense that each proposal, in its own different way, was envisaged to provide a single route by which taxonomy could address the twinned biodiversity *and* taxonomic crises, the two-pronged failure which was seen by this time, not just as a failure of science and policy but one of humanity writ large. The chapter describes how, for these extraordinary proposals to survive, they rapidly had to become and to be seen as 'ordinary' practices that might be adopted by the taxonomic collective. We chart below a delicate manoeuvring between extraordinary and ordinary claims often seen in accounts of scientific innovation, and we describe the ways in which one of the envisaged innovations – DNA barcoding – became an established part of the taxonomic landscape within just a decade.

The three proposals we highlight, from Charles Godfray in the UK (2002a, b and c; Godfray and Knapp 2004), Diethard Tautz *et al.* in Germany (2002 and 2003) and Paul Hebert *et al.* in Canada (2003a and b), each derive from distinct sub-communities within taxonomy and systematics. Each proposal aroused much comment and even controversy within the sub-communities constituting the broader field of taxonomy. Each of the three proponents for a *particular* kind of taxonomic 'progress' in the first decade of the 2000s put forward their ideas as a novel and perfectly logical 'route forward', which could have widespread *general* application for the discipline of taxonomy as a whole. Speed of identification and the speed and efficiency of taxonomic methods themselves were of the essence in all three proposals. The imagined new taxonomies would all in some way capitalise on the tools that the IT and genomic revolutions of the past decade had provided (McNally and Glasner 2007). These tools (bespoke bioinformatics software and information technologies including technologically advanced databases and automated data systems), would provide a number of advantages and would enrol many new institutions and actors: they would not only allow scientists to escape the current taxonomy crisis, they would also give new impetus to biodiversity research and policy.

Disagreements about the way forward for twenty-first-century taxonomy were not easily resolved in the mid-2000s, however, and as we shall see in subsequent chapters, such disagreements persist as we write. In the second half of the chapter we look at the reception of these unitary singular schemes by practising taxonomists, and in particular, by a group of taxonomists wedded to a much more *integrative* (rather than singular or unitary) approach to the taxonomic crisis. We close the chapter by noting that, as in many other areas of science, 'closure' on particular ways forward has been achieved 'despite the persistence of disagreement' (Lynch *et al.* 2008: 42).

Each proposal for novelty and discontinuity within the practices of taxonomy needed quickly to demonstrate that it was simultaneously 'continuous' with existing taxonomic practices. DNA barcoding was no exception here. However, we shall also see that Hebert *et al.*'s proposal (seen in 2003a and b), above and beyond both of the other two, was presented as 'the' radical answer to taxonomy's problems on the most ambitious, generalized and global scale. Hebert *et al.* (2003a and b) and, later, DNA barcoding supporters (e.g. Stoeckle 2003; Janzen 2004b; Janzen *et al.* 2005), present the boldest and most confident case for a 'new', user-friendly transformational taxonomy, based upon the prescribed global uptake of a standard methodology, a 'CO1 gene-segment identification system'. This system, it was proposed, would not only provide a 'reliable, cost effective and accessible solution to the current problem of species identification', it would also service taxonomy's deeper intellectual passions, generating 'important insight into the diversification of life and the rules of molecular evolution' (Hebert *et al.* 2003a: 313). Thus Hebert's proposal made gestures towards those taxonomists interested in understanding the evolutionary 'tree of life' and phylogeny as well as those working towards species delineation and description. Furthermore it was fervently believed that DNA barcoding had far wider potential in terms of social impact. It was imagined that this kind of taxonomy could 'improve the way the world relates to wild biodiversity' (Janzen *et al.* 2005: 1835). Thus, as we shall see, assumptions of DNA barcoding's future use, together with vastly expanded range of new users, were an integral part of these promissory proposals.

These were grand claims, going much further than already ambitious 'species-identification' preoccupations. In the second half of the chapter we describe some of the main disagreements that the proposed taxonomic visions engendered in the scientific literature and amongst taxonomists that we interviewed. Opposition was not, in large part, focused on the testability of proposals.[2] Many taxonomist opponents of the DNA barcoding vision, for example, were quite aware that the DNA barcoding hypothesis could never be fully tested – it could never 'confront the world in its entirety' (Hull 1998: 515). Rather, as most other studies of controversy in or through science have shown (e.g. Nelkin 1979; Thomson 2005; Lynch *et al.* 2008), opponents to DNA barcoding less directly challenged the credibility, trust and scientific authority of the proposed new system. The questions they began to ask reflected this: Were DNA barcoders capable of retaining the intellectual value of past taxonomic preoccupations even as they vigorously promoted certain necessities for the future (Mallet and Willmott 2003)? What were the wider epistemic consequences of this particular mode of taxonomic work and had DNA Barcoders considered these (Will *et al.* 2005)? What new taxonomic units might this new universal DNA barcoding method spawn, and what did such units mean for existing taxonomic units (Vogler and Monahan 2006)? And they asked questions about the reliability of this method in the wider social world: What might happen if non-taxonomists, who might not have the benefit of understanding of DNA barcoding's uncertainties, began routinely to use the results of DNA barcoding in biodiversity policy and decision making?

We end the chapter by describing how a way forward for DNA barcoding as a global standard for molecular species-naming has been navigated to date, without collective scientific assent. We describe how this came to be possible in the particular post-genomic context of the mid-2000s by virtue of a number of important strategic moves – including those of 'forgetting', 'naturalization', 'conciliation' and 'infrastructural inversion' – terms we explain later. In doing so we take care to avoid giving the impression that DNA barcoders were self-consciously manipulating the post-genomic context in such a way that DNA barcoding as a new kind of 'shallow genomics' could thrive. The kind of naturalization we witness in this chapter is consistent with attempts to stabilise new genomic technologies in other areas of the life sciences where a whole host of elements – technologies, regulations, epistemic norms, markets, users, benefits and so on – are emergent and in flux (Shorrett *et al.* 2003; Rabinow 1996; Fischer 2005; Franklin 2005; Thomspon

2005; Thacker 2005; Franklin and Lock 2003; McNally and Glasner 2007; Atkinson *et al*. 2009; Parry and Dupre 2010). Barcoding, like many other emergent genomic sciences, inevitably drew on the resources available to it in wider cultural terms, cultural idioms and ways of thinking – a new global genomic 'imaginary' was ushering in newly defined fields of agency and possibility (Appadurai 1996). The visions of DNA barcoders had to be global, and their ideas released them, and potentially all taxonomists along with them, from the smaller scale, traditional concerns of the taxonomic disciplines. Their visions were therefore bound to be controversial, transgressing the strictures and boundaries of taxonomy, connecting the identification of species up to an explicitly global technoscape (Appadurai 1996), freeing up thought in such a way that it became possible to think of the chosen CO1 microgenome as a planetary ordering device (Thacker 2005).

We describe here, in turn, Godfray's, Tautz *et al*.'s and Hebert *et al*.'s extraordinary proposals for a new science of taxonomy deemed appropriate for the twenty-first century.

Godfray's unitary taxonomy

Four articles published between 2002 and 2004 by Professor Charles Godfray, a population ecologist working at Imperial College London, exemplify in their own titles (including 'Towards taxonomy's glorious revolution!' and 'Taxonomy for the 21st century') a rallying cry to re-imagine the future of taxonomy and reinvigorate the discipline (Godfray 2002a, b and c; Godfray and Knapp 2004). Here, Godfray wrote energetically about the changes he saw necessary for taxonomy to 're-invent[s] itself as a twenty-first century information science' (Godfray 2002b: 17). On a superficial level he acknowledged that taxonomy had a dusty, tired and even out-dated image as a field. One worry was that, presenting itself thus, funding would be hard to come by. Godfray tackled this and deeper related issues head on: his proposal made dramatic reading, arguing that taxonomy as practised at the turn of the century needed a shake-up both in terms of organization and in terms of everyday practices (Godfray 2002a and b). The problem, as he saw it, was a lack of coherence: 'The taxonomy of a single group of organisms does not reside in a single publication, or a single institution, but instead in an ill-defined integral of the accumulated literature on that group' (2002b: 17). Godfray boldly offered, 'But this is not the only way to organize a taxonomy' (ibid.).

His first suggestion was that taxonomy needed to shift from being a science with a 'distributed' organization to becoming one that was 'unitary'. The many different sub-disciplines within taxonomy, which over the years had also come to be factions (Hull 1988) had to be brought together. This point had been argued some months before in *Antenna*, the Bulletin of the Royal Etymological Society (Claridge 2002), but Godfray's argument had an added twist. This would be achieved through the systematic revision of major groups of organisms to standards decided on by one existing international body. It would follow that the taxonomy of a particular group 'could reside in one place and be administered by a single organization. It could be self-contained and require reference to no other sources' (2002b: 18). His second point was that this unitary tax-onomy would need to make itself widely accessible through the World Wide Web. Lastly, he suggested, this web-based taxonomy would need to expand its epistemic boundaries – to encompass not only descriptions and evolution-ary relationships of living species, but also species biology and function. On the face of it, Godfray's vision involved a 'root and branch' shift in taxonomic epistemic and technical practices, including a serious commitment to unitary coherence achievable through the affordances of digital technologies and the World Wide Web.

In order to portray a radically new picture of the future, Godfray had to present a problematic picture of the past. Whilst appreciating the richness of taxonomic practices, and aiming to embrace both descriptive and phylogenetic approaches, he portrayed 'the legacy of more than 200 years of systematics' (2002b: 17) as a 'dead weight' for some taxonomic groups – pulling taxonomists backwards into the task of deciphering nineteenth-century systematics, deconstructing former publications and looking for specimens to re-examine in the light of current knowledge. Godfray envisaged a decisive break between these backward-looking practices and those possible in the future. Whilst careful to pave his way with caveats, 'I am not a pro-fessional taxonomist and am under no illusion that what follows will be the best or even a viable model, but I hope it will bring out the issues involved' (2002b: 18), he outlined a way of wiping the taxonomic slate clean from this burdensome history, as a way of freeing it for the great leap forward.

The clean slate (the necessary rupture and imagined liberation from the status quo), for Godfray, would be achievable through the implementation of a new formal taxonomic procedure. This procedure would be called the 'first web revi-sion'. As a kind of technology for creating change the first web revision would incorporate a delicate balance of continuation of past practices with a decisive orientation towards the new.[3]

> This would be a revision of a major group of organisms to a standard decided on by the International Commission on Zoological Nomenclature, or the International Botanical Congress, or equivalent body (let's just call it the international committee) … This draft first web revision would be placed on the web for comments from the community, then after changes have been made in response, it would become the unitary taxonomy of the group … [F]rom this time onwards all future work on the group need refer only to the set of species in the first web revision and then later to those in the 'nth (that is, current) web revision'. The taxonomy of the group is thus at a stroke liberated from nineteenth-century descriptions and potentially undiscovered synonyms.
>
> (Godfray 2002b: 18)

For those seeking its modern renewal, the idea of a clean slate for taxonomy – Toulmin (1990) had described for the Scientific Enlightenment the (perhaps functional) myth of the clean slate – was deeply appealing by the early 2000s. Many others, includ-ing taxonomists themselves, were also considering around this time how to abandon taxonomic histories that held them back and obstructed their ambitions and imagined future needs. But despite the vigour and possible violence of this unitary vision, dispensing with taxonomy's burdensome past 'at a stroke', Godfray in fact suggested that what he had described as necessary was an evolutionary rather than revolutionary move (2002b: 18). He believed that it would preserve the hard-won successes of current taxonomy while dispensing with its historical baggage. And in fact he recognized that this would not take place overnight and that, in practice, 'groups would move to the new unitary taxonomy as resources became available' (ibid.).

We see here the first of several instances of the 'Janus-faced' nature of claims making for a new taxonomy.[4] In the early 2000s the need for change coupled with equally important need for continuities in taxonomic practices had brought about a situation of tension. As everyone within the taxonomic community was well aware, the sense of crisis in taxonomy was aligned with the slow, idiosyncratic, low-technology, non-standardized and subjective images of morphological taxonomy. Talk of taxonomy's redemption, on the other hand, began to look more and more like a shiny new high tech, genomics-shaped, IT-rich, big, powerful and fast science, oriented to the policy needs of the coming century. Here we read Godfray as advocating a radical 'clean-slate' vision. Yet on the other hand, in the same articles, we read that his proposals allow for a gradual evolution of the field. We are witnessing here, and we shall see it also in the two other visions for the future of taxonomy in this chapter, the simultaneous use of discourses of discontinuity and continuity (Law 2002). In Godfray's terms, making discontinuities with taxonomy's past, a past which is characterized as a much fragmented and diverse tradition – is hailed as a strategic move, necessary to provide a much needed 'break' in which a new future 'unitary' system might emerge in a new cyber-intensive context (Hine 2008: 228).

But as analysts of such innovating moves, we remain alert to the tension between attempts to forge discontinuities and the actual impossibility of doing so (Suchman and Bishop 2000; Suchman 1994, 2007).[5] Of interest also are the strategies or mechanisms by which discontinuities are temporarily forged and continuity is suspended in order for a unitary project to gain momentum and traction. Discontinuity from past and present practices, as seen in Godfray's vision, and in those that we are about to document from Tautz *et al.* and Hebert *et al.*, implies a form of forgetting (and non-recognition of those 'forgotten') which was resented by others in the taxonomic community. It is also possible, here, to detect a form of what Bowker calls 'infrastructural inversion' (1994a and b). This lies in the sense that, while this process of strategically controlled forgetting may well be necessary for the establishment of new globalizing taxonomic projects, in the longer-term it will not hold water as a viable epistemic proposal for the whole of the taxonomic community. As Renan suggested, unity necessarily involves forgetting and is thus always brutally created (Renan 1882; Matsuda 1996: 206). Questions that need to be raised here are: Does the reductionism inherent in the effort to reinvent taxonomy inflict lasting 'damage'? Or will the 'new' taxonomy, in whatever form, sooner or later have to allow for mutual accommodations, and a related game of epistemic 'catch-up'? We come back to these issues in the second half of this chapter.

Hine suggests that the immediate attention that Godfray's proposals received was in no small part thanks to the 'package' of solutions he provided (in Godfray 2002a, b and c), just at the time when the UK Parliamentary House of Lords Select Committee inquiry was considering the future of taxonomy and when the UK government was considering its ability to meet its policy aims on biodiversity following the UN global Convention on Biodiversity (House of Lords Select Committee on Science and Technology 2002). Godfray had suggested that the way out of the taxonomic impediment – the inability of taxonomists to find, describe and classify the biodiversity of the world at a rate which exceeds the current rate of extinction of species – was seen (at least in the UK) as largely achievable by means of a cyber-oriented technical fix. He had advocated the use of information technologies to make a unitary, web-based and digitized taxonomy as the only sensible way forward. It was also, he suggested, the only way to make taxonomy 'fundable' at this stage in the discipline's deteriorating fortunes. He put his proposals starkly to the taxonomic community: 'there is everything to gain and little to lose' (2002b: 19). For the many in taxonomy and related scientific communities feeling their backs against the wall, web-based information technologies and fast gene-sequencing took on the status of highly 'evocative objects' (Hine 2008: 96).

A new scaffold: Tautz *et al.*'s DNA-based taxonomy

Whilst Godfray's vision for a viable global taxonomy for the future was established firmly around the use of web-based technologies, it only timidly introduced the case for a DNA-based taxonomy to play its role in the new scheme: Godfray did not envisage that the further challenge, of basing a future taxonomy around DNA sequencing, would become a reality for some 10 or 20 years (2002b: 19). Yet later that same year, building on Godfray's lead, European taxonomists were taking the opportunity to imagine just such a reality. Diethard Tautz *et al.* (2002, 2003) captured the energy and ambition of a group of researchers from Germany, Denmark, Italy and the UK who claimed that taxonomy should not only move towards a unitary web-based system but go further still: centre stage within that unitary system should be a DNA-based taxonomy, as the 'scaffold' around which the future of taxonomy globally could be based.

DNA taxonomy was already well established and had been for many years (Avise 2004, 2000). In some taxa, DNA sequences were already serving as the main way of linking voucher specimens to other information and were the primary referent and language for communication between experts. However, Tautz and colleagues' papers (2002 and 2003) effectively represented a proposal for DNA sequences to take on the role that Linnaean binomials had traditionally played in taxonomy. DNA sequences, in other words, would themselves serve as the taxonomic reference system across all groups (Tautz *et al.* 2003: 71; Vogler and Monahan 2006: 3).The radical nature of this proposal is, in effect, its tangible break with, yet subtle continuing reference of, the Linnaean system. As Vogler and Monahan reflect, the aim in DNA taxonomy is to identify groups that correspond to entities of reproductively coherent individuals (the species), i.e. to determine a hierarchical level roughly equivalent to the binomials of the traditional system (which are generally considered to represent the true species in nature). While recourse to the Linnaean nomenclature provides important evidence for the correct level of the taxonomic hierarchy, it does not follow that the DNA groups are only valid if they correspond precisely to the existing species names.

(Vogler and Monahan 2006: 3)

We single out Tautz *et al.*'s 2002 and 2003 papers, amongst many that promote an already established DNA taxonomy, for the way in which they *both* build upon Godfray's vision of a web-based taxonomy, *and* anticipate many of the universalizing ambitions of Hebert *et al.*'s CO1-based system for identification. Tautz *et al.*'s vision for a comprehensive 'scheme' in which the routine identification of specimens would take place through high throughput DNA sequencing facilities, would create the need for: new DNA storage facilities; new methods of curation; new roles for museums both as rich sources of samples for the new taxonomic methods, and as concentrated sites for new research; and new database and software development. As such, Tautz *et al.*'s approach would have much in common with Hebert's vision of a CO1-based barcoding identification system. Like Godfray's and Hebert *et al.*'s proposals, this scheme represented a whole package emphasizing comprehensiveness, ambition and scope:

The basic procedures of DNA taxonomy would be straightforward. A tissue sample is taken from a collected individual and DNA is extracted from this. This DNA serves as the reference sample from which one or several gene regions are amplified by PCR and sequenced. The resulting sequences are, as a first approximation, an identification tag for the species from which the respective individual was derived. This sequence is made available via appropriate databases, together with the species description and other associated information, ideally including its taxonomic status with appropriate references. The sequence now serves as a standard for future reference, together with the type specimen and the respective DNA preparation, which will be deposited in museum collections. Once a significant sequence database has been built up, new samples can be checked against these existing sequences to assist species re-identification or to assess whether a species description might be warranted.

(Tautz *et al.* 2003: 70)

DNA taxonomists envisaged this system as providing a DNA sequence database that would serve, when populated, for routine identification tasks. One sees here, whether deliberately promoted or not, a fundamental shift towards the weakening of the ontological role of morphological characters as differentiating markers for species definition. However, as we shall see in the next section, in explicit contrast to Hebert's barcoding proposals (Tautz *et al.* were quite aware of Hebert *et al.*'s 2003a paper, in press at the time of writing their own 2003 paper), there was no aim to standardize nor to automate methods any further than this – unlike DNA barcoding proposals which would be based solely around the promise of using a standard segment from a mitochondrial gene, CO1, DNA taxonomy could be based on one or more regions of mitochondrial DNA or nuclear DNA. It could be derived from clustering or from phylogenetic methods, using any gene region; indeed the sequencing of multiple genes was considered desirable (Tautz *et al.* 2003: 72; Vogler and Monahan 2006: 2). Thus, DNA taxonomy still required substantial skill and judgement from the user. Tautz *et al.*'s DNA-based scaffold would build on the strength of traditional systems whilst giving natural history museums new roles as molecular facilities and guardians of molecular and genomic diversity. The new system would be fast, economical and efficient. But above all, this was seen as a system that would usher in a new objectivity and universality to taxonomy:

> DNA sequence information is digital and not influenced by subjective assessments. It would be reproducible at any time and by any person, speaking any language. Hence, it would be a universal communication tool and resource for taxonomy, which can be linked to any kind of biological or biodiversity information … Although DNA taxonomy has limitations, it would have the advantage of being a universal tool.
>
> (Tautz *et al.* 2003: 71)

DNA taxonomy, although it already existed as a set of practices within taxonomy, was being repackaged here by Tautz *et al.* as part of an entire system that could enrol many other actors and institutions within it. This was a standardized package (Clark and Fujimura 1992; Fujimura 1987, 1988) that would define and bring about change. Essential in this were appeals to objectivity and universality within the science. Also crucial to this proposal was the idea of the hypothesis – the philosophical commitment to the testing of specific taxonomic theories, including the ever-elusive species concept (Tautz *et al.* 2003: 72; Vogler and Monahan 2006: 2).

Like Godfray, Tautz *et al.* stressed continuities: 'a DNA based system must be firmly anchored within the knowledge concepts, techniques and infrastructure of traditional taxonomy' (2003: 71). Their stance concerning the global standardization of DNA markers was relaxed and pragmatic: 'A universal agreement about the type of molecules and genes to be analysed does not seem to be necessary' (2003: 72). In their view, also, molecular and morphological knowledge could, and should, be fruitfully combined. Suggestions for the future of the taxonomic discipline were thus carefully crafted from an understanding of what needed to be preserved from past practices and philosophies of taxonomy, whilst at the same time presenting proposals for radical reform, extension and 'overhaul' of the present system and its practices.

Both Godfray's and Tautz *et al.*'s proposals had managed to put bold visions on the table whilst still claiming an 'evolutionary' or continuous approach with respect to existing taxonomic traditions. Proposals for DNA barcoding, also appearing in the literature in 2003, made perhaps the most audacious proposition for reform of the 'old' system, however. Riding on a wave of incredibly high ambition, Paul Hebert and colleagues at the University of Guelph, aimed to create a global bio-identification system for all animals and to extend the proposed approach to 'all life' (Hebert *et al.* 2003a: 313). This proposal, ultimately to DNA barcode all life on earth, through the idea that, for unique identification, *'one gene = all species = all life'* promised both 'a revolution in access to basic biological information and a newly detailed view of the origins of biological diversity' (2003a: 320).

DNA barcoding: a no-brainer for taxonomy?

Paul Hebert and colleagues (2003a and b) approached the perceived taxonomy crisis with an argument that was aimed directly at morphological taxonomy:

> Since few taxonomists can critically identify more that 0.01% of the estimated 10–15 million species (Hammond 1992; Hawksworth and Kalin-Arroyo 1995), a community of 15,000 taxonomists will be required, in perpetuity, to identify life if our reliance on morphological diagnosis is to be sustained.
>
> (Hebert *et al.* 2003a: 313)

Presenting their arguments as building upon the technologically framed vision of Godfray (2002), Tautz *et al.* (2002, 2003) and others (e.g. Blaxter 2003; Bisby *et al.* 2002), the researchers at Guelph did not skirt around what they perceived to be the glaringly obvious limits of the human capacity for morphological identification on a global scale. Arguments concerning the limits of morphological identification were combined with the painful reality of which all taxonomists were acutely aware, namely that the global pool of taxonomists was, in any case, dwindling rather than increasing. This paved the way for a seemingly simple proposal or a 'realistic solution' as it would come to be called (Costa and Carvalho 2007).

Hebert and his team of researchers at Guelph advocated a future for taxonomy that would be built not only on an increased technologization of the field, but on a very particular 'microgenomic' or 'DNA-based' identification system (2003a: 313; 2003b). Microgenomics implied fast and simple sequencing from a short 648 base-pair length of DNA alone. Identification was based upon the analysis of sequence diversity in this small segment of DNA called cytochrome oxidase subunit 1 (CO1). This kind of analysis was already accepted as a means of identification for morphologically intractable groups – organisms that are difficult to identify from their form alone, e.g. viruses (Allander *et al.* 2001) and bacteria (Hamels *et al.* 2001). Hebert and colleagues would portray their proposed new microgenomic system as a significant innovation in the context of the creeping progress of morphological taxonomy, however, and they went on to advocate and defend the creation of a DNA barcoding system for all animal life.[6] Whilst there were great hopes for CO1 in these early stages, it was recognized that it would be 'impossible for any mitochondrially based identification system to resolve fully the complexity of life' (Hebert *et al.* 2003a: 319). It was anticipated that the plant kingdom, in particular, would provide significant extra challenges because of the different evolutionary histories of plants and animals. Hebert suggested to us in interview in 2006:

> We know strange things are happening in CO1 in plants ... in some plant groups it's evolving like a rocket. And if it's slow in other plant groups this is where the interesting science starts to bubble ... When you're looking at the same gene region [CO1] across different kingdoms of life, you can now ask questions about what does it do in animals and what does it do in plants?[7]

Mixing cautionary rhetoric with an extremely ambitious plan, the Guelph team suggested in 2003 that 'Microgenomic identification systems, which permit life's discrimination through the analysis of a small segment of the genome, represent an extremely promising approach to the diagnosis of biological diversity' (Hebert *et al.* 2003a: 313). A more confident conclusion published later in the same year stated that 'CO1 analysis actually provides a taxonomic system that is chasing the last digit of animal diversity' (2003b: 99).

A CO1 'system', as Hebert *et al.* called it, coupled to a related database of known species could, it was suggested, be developed in 20 years to cover the 5–10 million animal species on the planet, all for around $1 billion US dollars – far less, it was argued, than the costs directed towards other big science projects such as the Human Genome Project or the International Space Station. As well as this predicted economy, adopting barcoding for the identification of animal species had a beguiling simplicity about it. The technology of the barcode, initially designed as a way of coding commercial products (the Universal Product Code, UPC) was seen to lend itself perfectly to the task of representing organisms through the short CO1 segment, only 648 base-pairs long (2003a). As reported in Hebert *et al.* (2003a) and on the website of the Canadian Center for DNA Barcoding at Guelph:

> Modern UPC symbols use an 11-digit series of lines, each representing one of 10 numerals, for a total of 10^{11} unique combinations. Each position in a DNA sequence has only four possible 'numerals', but a stretch of only 15 nucleotides nonetheless provides 4^{15} (>1 billion) possible combinations ... In contrast to the UPC system, in which a completely unique commercial barcode (which need only differ by one digit from other barcodes) is consciously assigned to each product, distinctive DNA barcodes are generated through the accumulation of random mutations between reproductively isolated groups of organisms. However, with a modest 2 per cent per million year rate of sequence evolution, a 600 base pair segment of DNA will, in theory, provide 12 diagnostic nucleotide differences between any two species that have been separated by only one million years.
>
> (Hebert *et al.* 2003a)[8]

As an aside, the incongruity, and ambiguity, of using commercial product bar-coding technologies to signal the diversity of endangered natural kinds was not lost on those who would later hear about DNA barcoding of life, including the artist Dion Laurnet, who proposed an art project that would follow the project to barcode all species on earth (http://www.dionlaurent.com/barcodes_of_life.htm; Larson 2011). We come back to this point in later chapters.

A crucial part of Hebert and colleagues' proposal was that this short gene sequence should be taken from a standard position on the mitochondrial gene, CO1. CO1, present in all animal species would therefore act as a stable pivot through which species differentiation could be detected. This requirement for standardization – of the organic reference-object in every global specimen collected and analysed – related directly to the ambition and the globalizing scope of Hebert's vision. In order for a reference system based upon the analysis of these sequences to work globally, CO1 needed to be a unique and universal barcoding marker, 'serving as a core of a global bioidentification system for animals' (2002: 313). Without such a standardized baseline reference as DNA marker, there would be no consistent basis for comparatively observable difference, and identification. The imagined universal use of the short and thus fast, simple and economical DNA fragment (CO1) that would enable a 'common registry' or database to be made, combined with CO1's assumed power in the diagnosis of every distinct animal species back into the existing Linnaean system, suggested an exciting set of possibilities for barcoding. The CO1 system would not only introduce a simple way of identifying species, it was thought that barcoding would be particularly useful where morphological methods were inadequate – in the identification of cryptic and polymorphic species, and where biological material was fragmented or deriving from life history stages (e.g. eggs) of an unknown source. Users would include policy, regulatory and health agencies where swift identifications might be needed and highly valued (Costa and Carvalho 2007). Furthermore, it would create order and meaning out of a field fragmented by different practices, material collections, methods and goals. It was imagined that the heterogeneous philosophies, materialities and practices of taxonomy would be realigned to this unifying system, itself based around a fragment of a gene, and translated into a genetic barcode that could be imagined to be 'embedded in every cell' of every organism (Hebert *et al.* 2003a: 313). In our interviews with scientists involved in the early days of barcoding, what was emphasized was the role of this gene fragment as an *organizing principle* – one which would enable global creation of, and access to, a biodiversity commons hitherto obscured by taxonomy's lack of organizational and intellectual unity.[9]

CO1 was hailed as an organizing principle in a new kind of 'horizontal genomics' that, unlike most genomic science, would not be oriented around understanding gene *function*, but rather around the identity-differentiation exhibited by a single gene segment held constant across all species. As such, this CO1-based ordering would not, however, supersede existing Linnaean taxonomy. Analysis of sequence similarity and difference of the CO1 gene segment across all known animal specimens would require the 'banking' of thousands of

CO1-derived barcodes in a new fit-for-purpose Barcoding of Life Data System (BOLD). But deposited barcode data would refer back to the samples and original voucher specimens already identified according to morphological techniques and Linnaean nomenclature. The 'barcode library', as this part of BOLD would come to be known, would then be used as a reference system against which unknown individuals could be matched. Unlike DNA taxonomy, therefore, the sequences derived from specimens are 'retro-fitted' to the Linnaean binomial classification system.[10] Hence DNA, in barcoding, is not the arbiter of groupings of living organisms, or the 'scaffold' of classification as Tautz *et al.* (2003) put it. It is simply a route – and a shorter DNA-sequencing route than typical DNA taxonomy would use – through which researchers can gain quicker access to established Linnaean binomial species groupings. The entire vision rested on the important assumption that variation, detectable via CO1 analysis, *within* Linnaean-defined species is usually lower than that *between different* species (Hebert *et al.* 2003b; Hebert and Gregory 2005; Vogler and Monahan 2006).[11] We look at this assumption and its implications in more depth in Chapter 3.

The logic of barcoding was thus portrayed as a way of harnessing 'the infrastructure of information indexing that's already established'.[12] Hebert *et al.*'s publications, 'Biological identifications through DNA barcodes' and 'Barcoding animal life: cytochrome c oxidase subunit 1 divergences among closely related species', published in the *Proceedings of the Royal Society*, in January and May 2003 respectively were shortly followed by the Banbury I and II meetings held at Cold Spring Harbour in March and September 2003, where ways forward for Taxonomy, DNA and BOLD were discussed between interested parties from the taxonomic, bioinformatic and funding communities.[13] DNA barcoding's 'selling points' were much vaunted, even though there were few scientific findings that could support the promise of CO1's *universal application at this time. The entire debate, organized around the 2003 papers and the Banbury and other conferences, was in a sense a kind of experiment in its own terms, and several questions – Does it work at all? Can the proposed universalization of method be carried out, scientifically? Do we know whether this DNA segment can represent *all* animal species? – were temporarily bracketed out in this early phase as the CO1 system was promoted through both scientific papers and at scientific meetings with possible future funders. It was hailed as a faster, cheaper, less labour-intensive, more practical, unifying and policy-relevant taxonomy than any previous system had ever been (Stoeckle 2003). The values of speed, cost, unity and relevance were of course just exactly those qualities that had been deemed lacking in the discipline, a lack that was also perceived as the main cause of the taxonomy crisis. It was also all these qualities, highly valued in all of the genomic, high throughput life sciences (Sunder Rajan 2006; Fortun 2009; Fischer 2005; Thacker 2005; Franklin 2005) which were seen to enable collective, globally coordinated mobilization of knowledge, towards an imagined policy effectiveness and utility. The sub-molecular material/informational pivot to all this epistemic and human possibility – the gene fragment, CO1 – was thus seen as an object that could transform taxonomy. As Franklin (2005) notes for stem cell technologies, a 'post-genomic, post-Darwinian, technique-led genotopia' was in the offing. CO1 seemed able to organize and bring together globally a multitude of new actors, institutions, technologies and aims whilst continuing to apply Linnaean names as a means of labelling and communicating about the diversity of living organisms.

Moreover, the envisaged system and database aiming for 'comprehensive taxonomic coverage of just a single gene' (Hebert *et al.* 2003a: 320) was hailed as an unprecedentedly powerful way of ensuring objective identification of species, allowing the identification of all life stages in animals, facilitating standardization hence comparability of taxonomic data, promising a revolution in humanity's *access* to basic biological information, hence improving taxonomic decision making *and* providing a 'newly detailed view of the origins of biological diversity'. Through the extraordinary capacities of microgenomic CO1 analysis, DNA barcoding, as Hebert *et al.*'s vision came to be called, was portrayed convincingly, especially to various funding agencies, as a kind of planetary bargain, a no-brainer for the future of taxonomy and for human capability to successfully identify biological diversity as fast, cheaply and efficiently as possible without sacrificing the intellectual goals of evolutionary taxonomy.

With these momentous promises for a taxonomy community wracked with anxieties of decline, fragmentation and failure, the further more difficult implications of this radical move were left in the background, at least for the time being. The more troubling dimensions of DNA barcoding of species were, however, picked up and pored over by other taxonomists, especially those who felt alienated by the singularity of the barcoding approach – its advocacy of CO1 as a standard gene segment and its relative emphasis on the identification services that taxonomy might provide to various new users. We look now at the reactions to Godfray's, Tautz *et al.*'s and Hebert *et al.*'s proposals.

The community reacts

> Changes in commitment made to particular theories, materials, instruments, techniques and skills in a laboratory, museum or in the field are not trivial matters. Although small changes occur continuously – for example, machines may break down, reagents may change, new technicians and students may be hired, funding may be lost and gained – the particular constellation of skills, tools, and approaches in each work site is not so easily or frequently changed.
>
> (Fujimura 1996: 10)

The ideas of Godfray, Tautz *et al.* and Hebert *et al.* built cumulatively on one another's suggestions in a very short period of time, and stood out from the more general taxonomic literature due to the unusual boldness of their vision, their global scope and their universalizing ambition. Each was portrayed as 'the' answer, an innovation and a grand vision, in response to the taxonomic and biodiversity crises. As all taxonomists recognized, the burst of ideas and debates seen in this part of the scientific literature from around 2002 onwards did not come from nowhere. Their articulation emerged from an ecology (Fujimura 1996) or dynamic assemblage (Deleuze and Guattari 1987; Law 2004; Ong and Collier 2005) of concepts, tools, practices, institutions, hypotheses, accepted and contested facts, problems, policies and imagined users and publics.

Three different approaches had been proposed that could be pivotal in bringing about wholesale and rapid change for taxonomy: the digital, web-based and unitary taxonomy of Godfray, the 'scaffold' DNA-based taxonomy of Tautz *et al.*, and the efficient, cheap, fast and accessible barcoding approach of Hebert *et al.* These three composite assemblages of different but overlapping specific aims, hopes, techniques, selected epistemic objects, questions, theoretical resources and networks, around which packages for change might be built, were carefully crafted to build in continuities with taxonomy's past, yet they also addressed several agendas of much greater scope and complexity. In its own terms, each was a proposed solution to the biodiversity and taxonomic crises combined, yet each resulted from the subtly different commitments to theory, philosophy, materialities and practices taken for granted in different sub-communities of taxonomic practice. However, certain common elements over-layered one another: if any one of the packages offered were to be taken up by the taxonomic community it would entail a technical fix (via IT, DNA or CO1) which itself would be bound up with the new databasing, infrastructural and DNA sequencing tools developed through genomics in the previous decade. Taxonomy looked set to join a whole raft of biological sciences that were taking advantages of these new genomics-related (including bioinformatics) tools. In so doing, they were beginning to absorb and reproduce the new norms and ideals of speed, cost, efficiency, molecular 'precision' and control, objectivity, universality, applicability and utility seen to characterise the post-genomic biosciences (Shorrett *et al.* 2003; Fischer 2005; Thacker 2005; Franklin 2005).

Taxonomists took at face value many of the grand claims of Godfray, Tautz, Hebert and colleagues, as well as claims made about the novelty of the proposed systems, and several publications and special issues appeared in the scientific literature after 2003 to address them. Critics began to debate the merits, or otherwise, of a 'unitary' web-based taxonomy (Mallet and Wilmott 2003; Scoble 2004; Hobern 2005; Godfray and Knapp 2004). Some parts of the community vigorously contested the idea that the current crisis in taxonomy could be relieved by DNA taxonomy and DNA barcoding (Mallet and Wilmott 2003; Seberg *et al.* 2003; Lipscomb *et al.* 2003). Other taxonomists lent support, promoting the use of DNA sequencing as a way to bring together phylogenetic research and species level taxonomy (Stoekle 2003; Oren 2004; Finlay 2004; Blaxter 2004). Taxonomists themselves described the reactions to these radical proposals as a furore, a tower of Babel, a cacophony (Mallet and Wilmott 2003; Wheeler 2004). Many warned that, unless something was produced out of this, unity of method or purpose for the taxonomic community would never be produced – the opposite danger seemed equally possible, that such vigorous debate might achieve yet further schism, more crisis, possibly chaos.

Within parts of the taxonomic community, an 'integrative' taxonomic approach to the taxonomic crisis was valued. Taxonomists such as Will, Kipling, Wheeler, Rubinoff, Meier and others felt that claims affirming the continuity of barcoding with established forms of taxonomic practice were 'thin', and the deletion of memory implied in barcoding's future-looking propositions was actually an arrogance. In emphasizing the need, perceived by some in the community as urgent, even desperate, to carry out the mundane and painstaking work of identifying species much faster and more efficiently than ever before, barcoding was accused by such taxonomists of pursuing only the most shallow of taxonomic goals (identification of *species*, or what taxonomists call the *leaves* or *tips* of the taxonomic 'tree of life') thereby evading the 'most important intellectual goals' of taxonomy – to understand the evolution of biodiversity or to generate and test hypotheses about species concepts (Wheeler 2004; Will and Rubinoff 2004; Will *et al.* 2005). What, from Hebert's point of view, looked like a creative and persuasive 'way forward' for taxonomy in the light of a global biodiversity crisis, for Will and colleagues, was seen as 'worse than bad', even 'destructive' of taxonomy's core purpose and goals (Will *et al.* 2005: 844). This apparent forgetting of core epistemic principles in the discipline led to charges that put the credibility of those proposing barcoding into question. 'Only through the ignorance of arrogance', wrote Will and colleagues, 'could one fail to learn the lessons of several centuries of comparative morphology' (Will *et al.* 2005: 846). Their point was that history matters. The question that followed was whether the future of taxonomy could be framed and led, on a global scale, by those who showed such little regard for the history and hard-won achievements of the discipline?

Infrastructural inversion and the making of new objects

What irked the community especially was the sense that the literature on DNA barcoding gave the false impression that it was Tautz *et al.* (2003) and Hebert *et al.* (2003a) who first proposed the use of DNA sequences for taxonomic purposes (Sperling 2003; Will *et al.* 2005; Meier, 2008: 97). Even though Hebert *et al.* had acknowledged some of their indebtedness to previous taxonomic theories and practices, these critics levied the charge that much of the novelty claimed for barcoding was misleading. DNA-based taxonomy had long been part of taxonomic practice, taxonomists had been using patterns and rates of nucleotide base-pair substitutions as characters with which to understand evolution and relationships between species since the 1950s (Avise 1994, 2004). The use of sequences of genetic data as part of the identification process had been pioneered since the 1950s. And even the idea of gathering a standard set of genes for identifying all taxa was itself not new (Caterino *et al.*, 2000). The use of microgenomic identification systems had gained acceptance with taxonomists working with morphologically intractable groups as far back as the early 1980s, and researchers had identified some of the attractive and useful attributes of the CO1 gene segment as a basis for this work in the 1990s (e.g. Folmer *et al.* 1994; Zhang and Hewitt 1997; Nanney 1982; Tautz *et al.* 2003). The use of a DNA-barcode for identification

(Hebert 2003a and b) was asserted to have been established previously (Moritz and Cicero 2004; Sperling 2003). The term 'DNA barcode' itself had been introduced ten years before Hebert *et al.*'s proposals (Arnot *et al.* 1993). Claims of novelty in the storage, retrieval and transmission of information (through the creation of globally accessible databases, etc.) were also refuted. The adoption of these technologies was seen simply as a natural adaptation of systematists to a rapidly changing IT context. Scientists, laboratories, funders and museums were already enrolled in the project of creating web-based taxonomic initiatives (e.g. All-Species http://www.allspecies.org; GBIF http://www.gbif.org; Species 2000 http://www.sp2000.org) or were beginning to be funded on a large scale (Godfray and Knapp 2004).[14] In a trenchant critique of barcoding proposals, Will *et al.* reported that 'critical assessment by any practising taxonomists quickly leads to a realization that what might be considered good in DNA barcoding is not new, and what is new is not good' (Will *et al.* 2005: 849).

What many critics realized was that, although the barcoding vision was portrayed as if it entailed an upheaval that would leave no part of taxonomy unchanged, in reality much of the mundane infrastructural work essential to such a new more positive departure was being or had already been done. The proposals represented, in this way, a classic example of the infrastructural inversion described by Bowker (1994a and b, 2005). Through infrastructural inversion – the making of grand visions and claims for action by DNA barcoders, once the hidden infrastructural work had already been done – extraordinary proposals come to be seen as materially sound and reasonable. At the same time, what are really continuities in practice can appear and be claimed as dramatic innovations.

Conciliation

What was 'new' was seen by some as an uninformed blunder, however, a blind and pre-emptive harvesting of DNA sequences, and the dangerous establishment of a new hybrid – the compromising 'gene-species' (Vogler and Monahan 2006), which in effect would be a Linnaean name, derived exclusively through CO1 analysis, rather than with or through morphological or other informed multiple character-based analysis. This 'gene-species' was an object that was said by critics to blend, in confusing ways, the open-ended possibilities of DNA taxonomy with the much more conservative epistemic boundaries of traditional Linnaean species nomenclature. Barcoding practice, that would only use CO1 sequence information to aid the study of curated voucher specimens, and to align these with previously established Linnaean names, was thus seen as a 'conciliatory approach', nothing more than a 'by-product' of the Linnaean classification system, rather than as a taxonomic system in itself (Vogler and Monahan 2006). Moreover for many in the community, this conciliatory approach meant a kind of 'fudging', perhaps even an evasion of scientific responsibility, since it amounted to a fundamentally un-ambitious system of Linnaean naming just at a time when taxonomy needed to demonstrate its true worth. According to some critics, what DNA barcoding evaded and shelved, precisely because of its *biological* lack of ambition, was the possibility that DNA-based information, even if derived from a single genetic marker, could contribute to the traditional intellectual goals of taxonomy – testing species concepts, questioning the biological nature of species, understanding the evolution of species and predicting species divergence.

These 'deeper' and – for many taxonomists reacting to barcoding – *truer* or defining intellectual goals, were only ever of secondary importance in papers and presentations given by Hebert and colleagues. As Paul Hebert suggested to one of us in interview, 'I've always been sort of more of the leaves on the tree-of-life kind of guy' – an approach that could be interpreted as more sensitive to the perceived urgency of the crisis of global biodiversity loss, aligned with a faith that barcoding was really going to help in mobilizing global citizens and institutions to protect biodiversity. As we have described, Hebert was busy, around 2003, trying to impress funders with the possibility of rapidly, economically, and even 'democratically' identifying these leaves or 'tips' – the living species occurring at the end of a long line of antecedents on every taxonomic tree. The possibility that DNA barcoding could quickly, cheaply and uniformly assign living specimens to agreed-upon species names across the globe was of far greater interest to barcoders than understanding branching and relationships between branches.

The latter emphasis – on relationships rather than names – was the principal interest of the so-called 'Tree-of-Lifers', a taxonomic sub-group whose interest was in understanding evolutionary relationships between species. Many 'Tree-of-Lifers' were concerned that DNA barcoding methods, as devised and persuasively promoted by Hebert *et al.*, would not be able to recover phylogenetically and biologically meaningful units. Worse still they might mask error by presenting an artificially simple view of the world, dressed in ostensibly innovative technology (Will *et al.* 2005: 847; Moritz and Cicero 2004: 1529; Lipscomb *et al.* 2003; Brower 2006; Kohler 2007). Barcoding, they worried, might well present an impoverished and misleading version of these tips and leaves, failing to recognize, pay heed to, or perhaps value, knowledge about the complexity of meaning and condensed evolutionary history contained in morphological characters. Furthermore, in suggesting that barcoding only applied to the leaves/tips of taxonomic trees, Hebert, they complain, was assuming two things: first that higher-level taxonomy would continue to be funded and carried out, as an essential service for barcoding; and second, that species are not themselves a phylogenetic hypothesis about relationships that are radically contingent (Will *et al.* 2005: 847). Thus, barcoding was argued to be presuming upon the proper scientific (hypothesis testing) research of others to work out *relationships* between species (the trees). Barcoding's own lack of depth; its urge to document comprehensively but only shallowly; its dependency on others to do the painstaking, drawn out science ('hypothesis testing'); and its simultaneous lack of heed to on-going debates about meaning; touched upon many sources of concern that were particularly sensitive within the taxonomic community, in the 'crisis' context of the early 2000s (Wilson 1999; Wheeler and Meier 2000; Will *et al.* 2005).

For Hebert, on the other hand, it was the critics of barcoding who did not appreciate the great potential of CO1 to re-orient taxonomy in the long-run. Those interested in the evolutionary questions of taxonomy, or the 'Tree-of-Lifers', as he put it, would come in time to see the power of the CO1 gene:

> When you inject every damn species on the planet you will be very surprised at the power you get when you put in just 1,500 different species … with five replicates of each and ask questions about their relationships. This single little gene region tells you quite a lot. It doesn't quite become an electron microscope but it's a hell of a lot better than what many people regard. And I think because it's going to be so cheap to do it and so comprehensive we're actually going to see it play a significant role in allowing people with tree of life predilections to decide what taxa they're going to work on. It's going to be the orienteering tool for tree-of-lifers.[15]

We can see here that Hebert's innovation, 'what was new', relied heavily on the extreme naturalization of 'this single little gene region', CO1, and its attested power in assigning unknown specimens to known species. CO1, indeed, seemed to offer an amazing constancy 'embedded' (as Hebert put it) in the living world. In Tree-Of-Lifers' terms, this was a role that was in some ways irrelevant since it in any case missed the mark in terms of taxonomy's deeper goals. Hebert and colleagues were not fazed by these challenges.

Taxonomic diversity and epistemic coherence

As we have suggested, some of the most vociferous critiques of DNA barcoding came from those taxonomists who advocated an integrated approach to the taxonomic crisis. Unlike the ethos of barcoding, this approach assumed the need for multiple discriminating knowledge-methods, each reflexively aware of its own limits, and thus of the mutual needs of such multiple paradigms, a complexity and interpretive incompletion which itself assumed skilled judgement on the part of any imagined user of the resultant science. Integration was thus assumed to be exercisable, perhaps flexibly according to context, by the discriminating user, as distinct from being a property 'hard-wired' into barcoding's 'black-boxed' automated system.

Hebert and colleagues' argument about the limits of morphological techniques and the need for an increased rate of species delimitation was not credible to the 'integrative taxonomy' community. According to them, the debate had been wrongly characterized by many as one of DNA-based versus morphological approaches. Against this, they argued that many kinds of taxonomy needed to use perhaps multiple DNA data, and many other types of data, such as on morphological characters, to delimit, discover and identify meaningful natural species and taxa. What was really at stake was the problematic proposal of a single character system (single gene systematics) that, through the reductive power of its vision and scope, would overthrow and disempower integrative, multiple-character systematics (Elias *et al.* 2007; Wheeler 2008; Meier 2008; Will *et al.* 2005). The weakness of any single-character method could not be overcome through further application or intensification of analysis of that one character, as with barcoding. According to integrative taxonomists, DNA barcoding had set itself a trap: stuck in a single character system, it had no way out – 'unless of course it brings in genes and morphological characters and becomes integrative taxonomy!' (Will *et al.* 2005: 846).

Facing the integrative community here was the sense that their own differentiated, situationally respectful epistemic practices were, almost by definition, non-coherent, non-controlling and for this reason non-seductive – in global 'public science' terms. Not only this, but perhaps also their continued defence of the heterogeneous epistemic traditions (e.g. in various branches of systematics, involving morphological character analysis) that had helped, by virtue of their continuing intellectual debates and disagreements, to cast the taxonomic community overall as non-modern, factional and unproductive (Hull 1988), made them appear fractured by dogmatic parochialism, and un-progressive. Integrative taxonomists thus witnessed with chagrin the seduction, by barcoding, of others from outside of the discipline,

> ecologists, behaviourists, conservationists biologists, etc., will, without a doubt, move ahead with items identified by DNA barcoding. They will accept the level of non-correspondence of these units to taxa and instead of taxa will use the so-called 'gene-species' … generating a false sense of security that nature has been successfully described.
>
> (Will *et al.* 2005: 846)

Will and colleagues further imagined that nature would thereby suffer: conservation agencies might use barcoding results in ignorance of the method and the hype around it and this could result in 'rash and irreversible mistakes that will impact significant elements of biodiversity' (Will *et al.* 2005: 848). Here, the relative value of the barcoding approach is judged by the putative consequences of (an assumed) *mis*understanding of it by others. Imagination of a falsely exaggerated value is thus evaluated as part of barcoding's dangerously seductive potential. As we shall explore more fully in Chapter 7, barcoders, however, had the ultimate trump card here: they could constantly refer to the immense task for which all taxonomists felt some responsibility – the need to inventorize all species on the planet, both as a route to their protection, and to avoid the horror of their extinction before even being known (Yusoff 2011).

Conclusion: the value and costs of a barcoding future

In what was one of the first responses to Hebert *et al.*'s 2003 papers, Mallet and Willmott, quoted at the beginning of this chapter, predicted that a revival of the discipline of taxonomy would inevitably also unsettle it. But as many histories show, upheavals were almost business-as-usual in taxonomy. Fierce arguments and debates about the fundamental questions and the methods by which these may be answered have never been absent in taxonomy: indeed, from early debates as far back as the natural history of Aristotle's time they have continued to characterize the field (Atran 1985; Lennox 1985; Hull 1988; Ridley 1986). Solid ground on which to base arguments for the proper identity of taxonomy has had to be crafted and re-crafted throughout the twentieth century as taxonomy seemingly constantly reviewed its definitions of itself (Hull 1988; Bowker 2005; Ridley 1986; Wheeler 2008). Arguably, no taken-for-granted core philosophy or set of practices in taxonomy actually existed in the early 2000s. In that sense, everything was up for grabs when DNA barcoding entered the scene.

Hebert *et al.* had declared in 2002 that they imagined it possible to create a database for all species on the planet within 20 years, for around one billion US dollars – a snip compared to the Human Genome Project or other similar big science initiative; and this, they asserted, would go towards building a science that had true relevance for nature, for nature-protection policies and for users on a global scale. This was seen by many, however, as a false, somewhat mystifying economy, neglecting the human costs of training and hiring of taxonomists upon which barcoding ultimately still depended. DNA barcoding's apparent gains were suggested to be possible only by circumventing necessary tasks and by creating a seductive but deficient product. Much data would be created, and fast at that, but at risk were their quality and value to science, and to society, as well as the understanding of what those data might mean. Though promoted as an answer to the problems of the overburdened and undermined taxonomic research community, the idea to barcode all species on earth was seen as adding extra pressures, as well as taking taxonomists away from some of their core values and goals, so far as these could be collectively identified.

In the scientific criticisms after barcoding's launch in 2003, there was a strong re assertion of the importance of the historical depth, scientific testability, scientific rigour, and the proper identity of taxonomy according to certain long-standing taxonomic principles. The rehearsal of these themes in animated arguments and debates between scientists within the taxonomic community at large demonstrates, we think, the extreme sensitivity of the taxonomic community to the attempts to re-write the history of taxonomy as a pre-requisite for establishing a route forward. These debates also show the extent of on-going controversy, bubbling away under the surface of what appears from the outside to be a single community, regarding taxonomy's principal goals and identity. Many of these deep questions and rifts in identity and purpose (and which of course go also to the heart of a taxonomist's daily epistemic practices) have long histories, and had hitherto been too difficult to resolve (Foucault 1996; Hull 1998). Wheeler (2008) recalls, for example, that attempts to make taxonomy simultaneously both a descriptive and an experimental science had historically ended in a confusion that had been papered over, only occasionally becoming the focus of explicit debate within the community, and never precisely resolved (Wheeler 2008; Eldredge and Cracraft 1980).

But it was not only the debates about taxonomy's history and the relative shallowness of barcoding efforts that stirred the communities of taxonomists and systematists to react. Barcoding heralded other new orientations within the science that we have so far only hinted at, and together these made this new venture in taxonomy profoundly, and necessarily, unsettling: an increased, intensified attention to users, including especially, non-specialist users; a commitment to global standardization and digitalization of data with all the changes in the material practical cultures of taxonomy that such a commitment implies; new relationships to funding, private foundations and big science funding sources; and the connection of taxonomic visions and ambitions to planetary biodiversity protection aims, and to a future in which our human effort to know nature precedes our dismay at its loss. As we have suggested, many of these commitments exhibit symptoms of wider transitions in knowledge cultures seen in other genomic sciences (McNally and Glasner 2007; Sunder-Rajan 2006; Fortun 2009; Atkinson *et al.* 2009; Franklin 2003, 2005; Fischer 2005; Thacker 2005).

Critics of barcoding did their best to question, fiercely in some cases, the credibility of barcoding as an answer to the biodiversity and taxonomic crises, also creating the space to put into question the scientific judgement of barcoding's proponents. Yet, as we shall see in the chapters that follow, it was still possible, even in the light of these vigorous challenges, to continue and to build the complex of skills, tools and infrastructure that would make barcoding a viable approach and set of practices that had universal scope and ambition into the future. Agreement within 'the community' was certainly not reached in the years immediately following Hebert *et al.*'s 2003 papers and concerns about barcoding are still being published in the scientific literature as we write. What brought about this apparent paradox – that a universalizing, standardizing, reductive, singular vision could go ahead without any real resolution of conflict and dissent within the taxonomic community – were certain devices or strategic moves that quickly translated the extraordinary proposals of barcoding into quite ordinary, inevitable and almost natural ongoing procedures. It is relevant that these were productive, in the 'normal science' sense that they were able to generate further technical questions and puzzles, and further feasible scientific work to address them. Such devices included those of: 'strategic forgetting' – using a particular selective rendition of the past to build arguments for a natural and inevitable future; 'naturalization' – the freighting of barcoding's innovatory potential upon the (strategically selected) natural attributes of CO1; 'conciliation' – combining the grandest claims for barcoding as the overthrowing of existing traditions, with apparently un-ambitious, already established mundane epistemic goals and values; and 'infrastructural inversion' – the making of 'revolutionary' and impressive claims for the future once the infrastructure for that future is in fact already in place.

Resolution of conflict between taxonomic practitioners in barcoding's case, it seems, was not a prerequisite for a programme of global standardization that was perhaps inevitably always going to be 'partial'. Our task in subsequent chapters is to look in further detail at what that 'partial universality' has come to look like in the decade since 2003, as the barcoding has been nourished, made material, disseminated, adopted, and adapted across the globe. Conflicts and frictions (Tsing

2005) are not absent from this story. We begin, in the next chapter, by describing how some of these frictions and resistances emerge from non-compliant natural forms themselves.

3 What's in a barcode?

The use, selection and (de)naturalization of genetic markers

In Chapter 2 we documented how the introduction of DNA barcoding brought with it a flurry of excitement and irritation within taxonomic circles. One potential of the phenomenon which few taxonomists could refute however was the power of DNA barcoding as an idea, which carried both rhetorical and practical force. The lure of the roll-call *one gene = all species = all life* proved potent due to its sheer simplicity. This had a range of interlocking ramifications. First, the DNA barcoding vision had the effect of seeming to reduce the immensity and complexity of the planet's biodiversity – something that is just 'too big for your brain' into a manageable and knowable array of natural organisms.[1] Second, as a renowned plant systematist at London's NHM stated, the mission to barcode all life using a universal identifier at least had the *potential* to provide the fractured global community of taxonomists with a unity they had lacked for 300 years.[2] Third, at the same time, BOLI researchers also promoted the idea of the DNA barcode as a much needed 'organizing principle' set to clean up and standardize the idiosyncratic data management practices of both taxonomy and the fecund, data-rich world of genomics.[3] Fourth, it was imagined that the simplicity of barcoding techniques and analysis meant that barcoding could be put to use in practical user domains such as pest control, the regulation of trafficked species, and ecological protection where life stages or body parts may require quick and reliable, repeat identification (DeSalle and Birstein 1996; Rubinoff *et al.* 2006). Fifth, the idea of barcoding species was regarded as something that might draw in mass popular appeal: barcoders imagined that the simple differentiation and representation of species, coupled to the Linnaean binomial system and graphically represented on the World Wide Web would have the effect of re-connecting human beings, all over the world, with a nature they had hitherto tended to neglect (Janzen 2004b).[4]

Claims about the extraordinary attributes of CO1, as natural and universal signifier of all animal species, emanated from the University of Guelph and were rapidly disseminated globally from 2003 onwards. However, these belied the extensive craftwork invested in the fabrication of this genetic marker.[5] As we show in this chapter and as seen in other areas of genomic science, such crafting involved a careful calibration between flesh, reagents, algorithms, sequencing and data management technologies. CO1 was not simply found as such in nature. Human agency and choice, work and negotiation had to take place for it to act as a universal genetic marker through which species differences could be established. The task of re-ordering single 'bounded' species (Helmreich 2009) as new material-semiotic objects involved dexterous technical work. The crafting of CO1 also involved skilful social and organizational calibration: it needed to articulate with visions of a useful, public, democratized science and with co-ordinating networks of laboratories that would take part in barcoding as a 'global standard in taxonomy'.[6] In this chapter we describe some of the biological, technological, organizational, political and human labour required to create and launch CO1 as the (sub-)genetic marker that would underpin the DNA barcode for global life.

Part of the successful emergence of CO1 relied upon its appearing as a natural voice – a ventriloquist, as Haraway puts it – for species identity (Haraway 1997). We thus locate our account of CO1's success in the first half of this chapter within a broader STS interest about the ways in which biological organisms and technological artefacts come to be endorsed through processes of construction, then naturalization. This involves an implicit reductionism, which we shall also unpack. Having demonstrated the ease with which CO1 came to be seen as the natural key to all animal species identification, in the second half of the chapter we turn to the complex set of procedures required to test and select adequate markers for plant species. We describe here how CO1's projected status as a simple, 'natural' and universalizing tool able to bring life's order and the global public together through a new form of taxonomy, was disappointingly short-lived. As botanists joined the barcoding race, what Hebert and colleagues had already hinted at (2003a and b) quickly became apparent: the extraordinary attributes of CO1 for the identification of animals were not realisable in plants.[7] As we will explain in more technical detail below (and also see the Technical interlude), CO1 simply did not contain enough base-pair variability to discriminate plant species in the strikingly effective and complete way it had appeared to do for animals.

By 2005 the search for a genetic marker for plants was fraught with the concern that the initial promise to barcode *all life* was under serious threat. At stake here was the promise that barcoding, as a new and faster form of 'shallow genomics', could be carried out to aid in the 'urgent' identification of all species on the planet. We note here that taxonomists working with fungi and protists also felt the pressure emanating from barcoding's upbeat promotion, to come up with standardized genetic regions for barcoding. *All but* the higher animal species, in fact, were the subject of much discussion, special working groups and attempts to identify standard markers, under the auspices of CBOL (the governing infrastructure for BOLI).[8] What was particularly interesting about this process was the contrast between the apparently seamless naturalization of the *animal barcode*, on the one hand, and the exposure, on the other, of the many contingencies integral to fabricating, testing and finally selecting alternative genetic markers for the less obedient natural kingdoms of plants, fungi and protists. Such contingencies, as we show, had the potential to disrupt the very essence of life's standardization as promoted by BOLI up to this point. The search for plant DNA regions, in particular, paradoxically exposed the artefactual qualities of CO1. It also worked to reveal, and render more widely visible, the attempted naturalization and hegemony of

BOLI through the much heralded success of DNA barcoding as a technology that could reveal all of life's diversity. By rendering transparent the work needed to naturalise CO1, the attempt to establish the plant barcode made visible the hubris or 'folly' that we referred to with Daston's use of the Babel metaphor in Chapter 1 – the limitations of a human endeavour to find a stable unchanging reference point encompassing and identifying the diversity of all life on the planet. However, we argue that the very process of searching for and constructing the plant barcode created the potential for a more robust, self-reflexive and creative BOLI.

Appreciating the artifice of Cytochrome Oxidase 1

> Genomic approaches to taxon diagnosis exploit diversity among DNA sequences to identify organisms … 'genetic barcodes' are embedded in every cell.
>
> (Hebert *et al.* 2003a: 313)

Molecular taxonomy is a field that creates intimate co-dependencies between flesh, code and informatics – fertile relationships that have increasingly come to characterize genomic approaches to the life sciences. These relationships require, however, that selection of the 'right' gene region for species iden-tification across different species takes into consideration the characteristics of that region. To put it another way, different gene regions offer the tax-onomist different possibilities for universal comparison and differentiation of species. This is true even though, as explained in Chapter 2, DNA barcoding does not enter into questions of gene function, but only into those of molecular sequence. In this sense it is akin to the (genome-)mapping ethos, as distinct from (functionally focused) hypothesis testing. The *sine qua non* for the bar-coding technique is that a short segment of DNA, universally present in all species, must nonetheless exhibit robust species-specific sequence variation (Hebert *et al.* 2003a and b). What underlies this apparent variation is rapid and frequent enough nucleotide substitution (ACGT), as we have outlined in the Technical interlude. However, as with any 'salient feature' selected in tax-onomy as the basis for 'splitting' or 'lumping' species, it is the possibility of detecting variation within it that is essential, thus the gene regions that offer this become salient features that make barcoding, and the entire concept of barcoding as a globally standardized method, work. However, such regions must also be considered together with more practical factors concerning the ease, speed, cost and reliability with which they can be retrieved, aligned and sequenced. Different regions of the genome are different in these practically important respects. Indeed, for the global barcoding endeavour that BOLI rep-resents, this list of requirements extends almost indefinitely – the gene region has to be compatible with the demands of global standardization, across all species, and across taxonomic cultures, so as genuinely to act as the pivot around which a new, publicly accessible, global barcoding effort might be real-ized (see the Technical interlude for the list of CO1 attributes that make it compatible with these demands, for the animal kingdom).

In the early 2000s Hebert and his team had many resources at their disposal for their task to select 'the' gene region for barcoding life. Several decades of research in molecular taxonomy had accumulated a depth of know-how on the advantages and disadvantages of different gene regions for a range of applications. In addition, the selection of particular gene regions for diagnosing species identity could be aided by referring to a reservoir of banked DNA sequences stored in the publicly accessible database GenBank.[9] The search for a DNA barcode therefore took place entirely within the context of the known – the prior establishment, via GenBank and other databases, of the links between genes, gene regions and species identity. As we have already suggested, the novelty of barcoding lay not in the science itself, but in testing whether a standard gene segment, could, on a massive scale, accurately and consistently discriminate between species. Thus, barcoding's 'structural' biological concerns, albeit extremely demanding in terms of global technical logistics and coordination, did not have to enter into functional questions. This was 'shallow genomics' *par excellence.*

In fact, Hebert did not spend long experimenting with different genes – after carefully sifting through the evidence, he quickly suspected that a 648 base-pair section of the cytochrome oxidase 1 gene – a mitochondrial protein-coding gene – was the best candidate for the job. Before we list and explain the various positive attributes of CO1,[10] the following excerpt from an interview in 2006 with Paul Hebert gives a sense of the way he has creatively grappled with many factors – biological, social, economic and technological – in pursuing the search for a universal marker:

> I began to play around with various bits of DNA and looked at a lot of the standard gene targets that people examine – the ribosomal genes, for example. If you look at nuclear genes you soon get reminded of a fact that all evolutionary biologists know – a lot of it evolves really, really slowly and it's not very effective at telling species apart so you've got to look at a lot of DNA to delineate species. By contrast, mitochondrial DNA is really easy to recover and you've got a finite number of targets and, having played around with different genes in the mitochondria in different groups of organisms, I began to develop an understanding of what bits of mitochondrial DNA were most easily liberated. And the other thing that has been a motivation to me from the time that this enterprise began was to try to find a system that anyone could use – that could be automatable. You know tasks like alignment can consume immense amounts of time, employing secondary structure models trying to align ribosomal DNA . . . I didn't want to do anything complicated, I wanted simple. And when you go for an interior part of a protein-coding gene like COI it turns out that the structure of those proteins are so conserved that it is extremely easy to automate the alignment. There's just no ambiguity in this task.[11]

Having sold the merits of CO1 on the grounds of its unambiguous signalling of species identity through easy extraction, and easy alignment, Hebert then goes on to reveal – in a slightly more reflective tone – one of the necessary trade-offs at the heart of the concept and practice of DNA barcoding:

> And so it was, you know, recognizing things that other people have recognized – that you can tell species apart with DNA, but then pushing it a little bit harder and saying not all bits of DNA are made equal, even within the mitochondrial DNA. Backing away from the single minded pursuit of the most variable piece of DNA (because a lot of people are fascinated by the fact that – like go into the 'D loop', I'll be able to find more variation here than there but you won't be able to align it, but there will be all sorts of problems when you try and compare across groups …). So, I was relaxing a bit on the resolution front. I was being very hardnosed about the ease of data analysis, very hardnosed about recovering the target piece of DNA, because CO1 is actually the slowest evolving of the 13 proteins in the mitochondrial genome (which seems paradoxical if you want to tell a species apart). But if you think about it for a lot you realise there's going to be lots of variation. And as soon as you move to a phenetic[12] approach, that's part of the simplicity argument: make it accessible to people as opposed to making it accessible to the practitioners of phylogenetics![13]

Whilst 'relaxing a bit on the resolution front', Hebert vividly portrays how technological constraints which were also social, crucially influenced the selection of CO1 as *the* genetic region that could underpin the universalizing and standardizing goals of DNA barcoding. These constraints included: the use of primers[14] to isolate and retrieve the selected gene segment before replication; the choice of software used to align and thus read genetic sequences; and a considered imagination of future users to ensure this technology was simple and accessible enough for a wide range of 'user-publics', such as pest-control officials, regulators of species trade, or even children (Ellis, Waterton and Wynne 2010).

The bottom line in the choice of CO1 is therefore not its distinctly biologically defined tendency towards frequent nucleotide substitution and thus its quality as an appropriate biological identifier, but its ability to *articulate* (Clarke and Fujimura 1992; Fujimura 1987), or as Suchman would suggest, its ability to *align* with a heterogeneous mix of other components, themselves contingent and in flux (Suchman 2000). This is a reminder of the fact that the knowledge constructed by molecular biology is itself a complexly constellated artefact (Katz 1989), the creation of which must enrol – which also means respecting and acting with, in some dimension – organisms, nucleotides, genes, primers, automated PCR machines, sequencing software, biologists of different specialties, users and imagined publics. The work, and the achievement of Hebert and his team, was to carve the most appropriate gene region, CO1, out of all the conflicting variables in play. In interview, this crafting, artifice and contingency is quite apparent. However, in 2003, CO1 was pronounced, like many other genomic entities, to be a naturally occurring object 'embedded in every cell'. Its naturalness, and its separation from discourses of artifice and nurture in fact had to be explicitly asserted to sustain its 'expansionist claims' as we shall see below (Keller 2008; Franklin *et al.* 2000; Franklin 2003; Davies 2010; Waterton 2010).

CO1 as universal standard

The task of universalizing CO1 as a standard requires that the methodology for its production is also standardized and universalized. Part of what CO1 had to provide was an ability to resolve species definition from samples prepared with cheap and easily accessible universal primers, using off-the-shelf algorithms for the definition and detection of similarity and difference, and using a process that could be easily understood through standardized protocols. Hebert and his Guelph team did not want to start off a series of global experiments to achieve these things: they wanted to be able to present a standardized package (Clarke and Fujimura 1992) to taxonomists and others all over the world, an off-the-shelf product for global taxonomy.

By 2003, Hebert and colleagues could report that they had carried out a study to test the feasibility of using CO1 for the accurate identification of animal species. Firstly, they had created what they call CO1 'profiles' at three different taxonomic levels (phyla, orders, species).[15] Here they had used known specimen DNA and investigated whether the sequences could be assigned to the correct *phylum, order and family* using only a CO1 derived DNA barcode (rather than longer sections of multiply composed DNA). They had found CO1 was 'sufficient to reliably place species into *higher taxonomic categories* (from phyla to orders)' (2003a, our emphasis). They had also found that CO1 nucleotide sequences could be used reliably to differentiate between different known *species*. On this basis, they concluded that CO1 analysis could resolve species boundaries, and furthermore, they suggested that a CO1-based identification system could be developed for all animals. The second study (2003b), again using existing sequence information from Genbank, had tested the ability of CO1 to discriminate between congeneric (similar) *taxa* in the major animal phyla. The results were good: more than 98 per cent of species pairs showed that CO1 nucleotide sequence divergence was within what had been decided then as a two per cent threshold for acceptable difference between aligned sequences.[16] Hebert *et al.* concluded that *species-level diagnoses* could 'routinely' be obtained through CO1 analysis. Looking optimistically at their results in terms of species identification, they suggested that

> Even if CO1 analysis simply generated robust generic assignments, its application would winnow the 10 million animal species down to a generic assemblage averaging less than 10 species, delivering a resolution of 99.9999 per cent of animal diversity in the process.
>
> (Hebert *et al.* 2003b: 99)

Getting closure on that 'last digit of animal diversity' was a goal that barcoders imagined, in 2003, to be almost within reach.[17]

As we saw in Chapter 2, reactions to barcoding claims were initially very varied. Whereas in Chapter 2 we rehearsed a range of arguments focused around barcoding writ large, here we focus on reactions concerning CO1 itself. The selection of just one reference gene, although conventional and of proven value in microbial taxonomy and in large-scale phylogenetic analyses, was seen, by some, as a problem. The big question was whether 'one gene [and a *mito-chondrial gene* at that] fits all' (Moritz and Cicero 2004: 1529). The potential limitations of using mitochondrial DNA (mtDNA), which allows tracing of gene flow through maternal inheritance, to infer species boundaries were seen to be many. These included: male-based gene flow; retention of ancestral polymor-phism; selection on any mtDNA nucleotide; introgression of mtDNA following hybridization; transfer of mtDNA gene copies to the nucleus; and the low pre-cision with which coalescence of mtDNA predicts phylogenetic divergence in nuclear genes. An additional problem was thought to be that focusing on mtDNA divergence would lead taxonomists to overlook new or rapidly diverged species, thus giving the false impression that speciation requires long-term isolation. All of these problems, according to Moritz and Cicero, were well documented in the literature (Bensasson *et al.* 2001; Ballard and Whitlock 2004; Hudson and Turelli 2003; Moritz and Cicero 2004: 1530) and were acknowledged by Hebert *et al.* (2004). Many of them concerned the ability of CO1 accurately to indicate species difference. However, they also revealed the necessarily close connec-tions between species identification and genetic history, which was a sensitive issue, and a problematic relationship for barcoders, who were trying to escape the long and seemingly never-ending debates about evolutionary relationships and phylogeny. Critics of barcoding could not accept simply abandoning this relationship, as we have discussed in Chapter 2. The problem as such critics saw it was that the history and contingencies of evolutionary relationships have direct consequences for species delimitations: evolution and speciation are inti-mately related! On the basis of these points, Moritz and Cicero contended that mtDNA divergence is 'neither necessary nor sufficient as a criterion for delimit-ing species'. Like other critics (e.g. Will *et al.* 2005, see Chapter 2), they urged that species delimitation should be based on 'multiple lines of biological evi-dence' including morphological, behavioural and functional, as well as genetic data (Moritz and Cicero 2004: 1530).

At the same time as responding to critics through the literature,[18] Hebert crucially presented his findings to live audiences. Several of these face-to-face meetings proved to be of pivotal importance for the future of barcoding. At what are referred to as the 'Banbury I and II meetings' held at Cold Spring Harbour in March and September 2003, for example, ways forward for taxonomy, DNA taxonomy, and the Barcode of Life Initiative (BOLI) were discussed, not only between taxono-mists, but between a mix of taxonomic, bioinformatics and – perhaps most crucially – funding communities. The director of the Program for the Human Environment within the Sloan Foundation had first been introduced to barcoding when attending one of Hebert's earlier presentations (prior to the 2003 publication) in Nova Scotia at a meeting organized by the Canadian Centre for Marine Biodiversity in 2002, but his attendance at the Banbury Meetings was vital. Here, Hebert's promotion of CO1 as the key to the barcoding of life was reported to be utterly convincing. Subsequently, the Sloan Foundation, together with the Moore Foundation, provided the backbone funding for major barcoding initiatives, including CBOL, without which, it is generally agreed, the rapid global expansion of barcoding would not have been possible. The Sloan and Moore Foundations already had a strong record of supporting visionary projects aimed at archiving and protecting biodiversity on a planetary-scale, such as the Census of Marine Life. The role of their funds, and their associated interests, was recognized by all participants as being of primary importance in creating and shaping BOLI as it is today.[19]

The Banbury I and II meetings were crucial for the coming into being of BOLI and for the establishment of CBOL and other supporting structures; but more immediately, they were also crucial for the naturalizing of CO1 as a universal marker. As the sociologist Barry Barnes suggested (2010: 220–221), naturalization, or 'essentialism' as he called it, requires careful observation on the part of the sociologist. What requires attention is not whether plausible explanations can or can not be given for the successful functioning of essentialized 'genocentric' or biological entities. What really requires attention is to understand why naturalization occurs, in order that we can understand the *work* that naturalization *does*, once achieved. Essentialism, Barnes reminds us, makes particular, controlled, assertions about the content of the world. But this has a function. It is a 'shared strategy' suggests Barnes, 'to facilitate selectivity of a given sort, to encourage forgetfulness, and thereby to facilitate coordination among a community of researchers' (ibid.: 221). To put it another way, we might see this process as a way of 'trying-on' a particular claim about the world – asserted as given and natural, but importantly, open to rebuttal and qualification. In the possible absence of any such rebuttal, however, the chosen objects and thier chosen qualities come to be accepted as given, and as natural. Hence, when Meier (2008) observed that taken together the evidence in the publications, *and* the refined performance and networking skills of Paul Hebert, combined to propel the motif and science of CO1 *beyond the internal struggles* of the systematics community (who at this time were still reeling from the publication of the articles), we can see one way in which a naturalized CO1 was skilfully being put to work by Hebert and colleagues. Barcoding was aiming beyond taxonomy alone.

Within the systematics community, Hebert's globalizing ambition and his drive to portray CO1 as the key to the barcoding of life was ill-understood. As we discussed in Chapter 2, general consensus about the way forward for taxonomy at such globalized levels was virtually impossible for the fractured community to reach, and the most comfortable resting place was at the time presented as a tepid, but perhaps fertile middle ground exemplified by calls for an 'integrative' taxonomy that locates DNA barcoding more modestly alongside other molecular but also morphological methods for species delineation and identification (Will *et al.* 2005; Moritz and Cicero 2004). This more complex and multivalent position, however, could not possibly have provided the rhetorical thrust necessary to access funding and convert taxonomy into a global, unified and *mobilizing* science. Paul Hebert understood this point extremely well. Lukewarm or circumspect proposals tend to be 'weak enrollers', stay local, and never gain global traction and support. Hebert on the other hand, knew he needed a bite-sized, potentially glamorous, hard-nosed and uncompromising proposal. Whilst most taxonomists were hedging their bets by acknowledging the benefits of a combination of molecular and morphological taxonomy, Hebert advocated the adoption, by all, of a single gene sequence for universal animal (and perhaps more complete) species identification, thus gaining enormous momentum and support from people and institutions in the right places. This was exciting: Hebert and global DNA barcoding with CO1 was on a roll. He even claimed at times not to be doing taxonomy! All he was advocating was a cheap and credible (fully standardized) tool to transcend the complex intellectual problematics that had traditionally slowed species identification right down, almost to a grinding and self-defeating halt.

What becomes clear if we pursue this point further (that what Hebert was proposing was not really taxonomy) is the extent to which Hebert *et al.* were drawing on a quintessentially genomic paradigm that requires a certain degree of naturalization in order to enable the instrumentalization of genetic materials (Franklin 2003). Furthermore, although it is acknowledged that genomics is an evolving field and by no means a 'stable referent' (Sunder Rajan 2006), its meaning and significance is seen to derive from more than technological innovation or epistemic advances alone: genomics, as Sunder Rajan suggests, 'has also been conditioned, in significant measure, by what is deemed a potentially successful business model at the time' (2006: 28). Genomics 'represents the rapid, high volume analysis of information, what is known as *high throughput* science' (ibid. italics in original). As such, naturalization is required as part of a massive process of standardization, which involves the attraction of broad allegiances, influential partners and large amounts of funding (McNally and Glasner 2007; Atkinson *et al.* 2009; Fortun 2009; Cooper 2008; Sunder Rajan 2006). Implicit here is the point that the scale of ambition and promise is roughly proportionate to the intensity of pressure for naturalization of key technical constructs, in this case the CO1 standard.

Here it is also relevant that as a genomic science barcoding enjoyed a wider range of reference points, with the momentous cultural shift or reconfiguration that was occurring in the life sciences as a whole (Rose and Rose 2012). In this larger canvas, taxonomy's own internal strife was a parochial affair. Like other life sciences, barcoding was struggling to shift, for example, 'from investigations within small laboratories to large-scale initiatives made possible by the establishment of extended collaborative networks' (Cambrosio 2009: 465). The novel collaboration between Hebert's group and the accomplished global scientific networking capacities of CBOL at the Smithsonian Institute in Washington DC, which was supported by the Sloan and Moore Foundations, became an effective organizational as well as technical vehicle for this more ambitiously intensive big-science biology, orchestrating a developing global hybrid technoscientific enterprise. Our point here is that the work done by Hebert and colleagues in the 2003a and 2003b (and subsequent) publications was ground-clearing: the creation of a simple iconic claim to provide him with a currency for stability and credibility during his global walkabout. This stable, black-boxed iconic new 'natural' tool was required in order to elevate barcoding into a fully fledged powerful *genomic* science. As such, Hebert was highly aware that CO1 was an essential object around which a genomic 'plan' could be made. As McKenzie (1989) has described with his notion of the 'uncertainty trough', the various contingencies and qualifications we have described above to the idea of CO1 as a universally effective identifier were not apparent and may not have been interesting to its various adjacent scientific disciplinary communities. They were sophisticated specialist 'users' of, or 'providers' for barcoding, and they had their own uncertainties and contingencies with which to contend. However, what Hebert also anticipated was the need for this plan to become, without too many hurdles, 'situated action' (Suchman 2007). Thus, a large part of his work was to portray CO1 as the 'natural' universal cog that would bring with it unbounded promise – of democracy, speed, power, economy and a frictionless ease – a promise that would appeal with equal success to a plethora of varied socio-technical situations. Through highly skilful manoeuvring, and through CO1's naturalized status as species identifier for higher animal species, barcoding gathered momentum to transcend and travel far above and beyond the quagmire of scientific debate. As it did so it would gain a kinship with many of the life sciences that now collectively lead in the giddy, promissory, speedy, global, genome-oriented bio-knowledge economy.

Simplicity and complexity in genomic science

What we have described so far as the process of selecting and stabilizing CO1 as the basis of the DNA barcode began with Hebert *et al.*'s claim that '"genetic barcodes" are embedded in every cell' (Hebert *et al.* 2003a). Of course, no such barcode itself exists in *any* cell – this is the interjection of a metaphoric reference, pregnant with other meaning (Larson 2011) into the known properties of the mtDNA within 'every' cell. And the message that flamboyantly (and somewhat hopefully) followed was that *one sequence = one barcode = all life*. In essence, what Hebert and colleagues were declaring here was that the crucial genetic constructions of life itself are already innately standardized; they are so regular, so stable, simple, and in a sense, so predictable that they can support a universal, standard, form of analysis. Furthermore, the ease with which CO1 complied with the ambitions of Hebert's team not only smoothed the way for a gradual acceptance and uptake of the DNA barcode but also apparently confirmed Hebert's initial assertion of the *naturalness of CO1* as a standard for knowing and taxonomically organizing all animal life. As Paul Hebert remarked in interview, 'I wouldn't have thought the world was so simple but it is!'[20]

Such tales of simplicity, whilst alluring, tend to arouse interest and suspicion amongst STS scholars, perhaps especially those who have scrutinized the powerful and on-going mythology of 'the gene' (Keller 2000; Kay 2000; Nelkin and Lindee 2004; Haraway 2007; Stotz *et al.* 2004; Franklin 2003; Wynne 2005). Simplicity is often suspected here of belying a hidden complexity. But STS scholarship on more mundane issues such as standardized technoscientific infrastructures, and the production of large, interoperable data sets, has also shown the essential reliance that these place, for their smooth operation, upon large bodies of hidden work and discipline, often involving 'simplification', 'naturalization', and 'standardization' (Bowker 2005; Hine 2006; Ong and Collier 2005). We note here that what was required to launch CO1 as a *standard* by which to judge life's diversity relied upon two essential factors. First, as in other examples from the life sciences, the transition from flesh-to-code-to-digital information, as part of the process of DNA barcoding, required close collaboration between systematists, laboratory managers, molecular 'wet-lab' scientists, laboratory technicians, bioinformaticians and a range of (off-the-shelf) primers and algorithms necessary for CO1 sequencing, representation and storage. The potential for these finely tuned collaborations to go wrong, or to diverge from their initial aims, for any particular reason, was effectively contained by the ease with which CO1 worked. Its success as a faithful, easy and cheap species indicator meant that few technical 'dead-ends' were travelled down, and many possible alternatives were left undisturbed or unexplored. Second, what appears as a simple, natural object – a feat of extraordinary and determined reductionism in the face of human, organizational, biological and technological complexities – is in fact a carefully constructed and finely tuned artefact, involving not only biological and technological agency and labour, but also as mobilizing agent, the vital articulation of a wider democratic and even salvationary, normative planetary vision from the taxonomic and broader funding communities (see Chapter 7).

The co-dependencies described, between a necessary fiction of (natural) simplicity on the one hand, and hidden epistemic–technical–political complexity, on the other, resonates with parallel examples from the life sciences and beyond. We note here the wider resonances with human endeavours to understand biological systems with human and planetary therapeutic futures in mind. The Human Genome Project, for example, provides similar insights, and was in itself an inspirational force for BOLI, and so we turn briefly to a consideration of this wider syndrome of the creation of simple and complex artefact in genomics.

Part of the initial power of the HGP rested on the hope (and promise) of the already-established DNA central dogma, that *one gene = one protein = one behavioural trait*. The sequencing of the entire human genome, it was imagined, would thus pave the way for major innovations in health, including the development of personalized gene therapies (Rose 2007). The therapeutic and other futures for human bodies promised by the HGP and its attendant genetic determinism reminds us of what is imagined for planetary biodiversity (a therapeutic future for all non-human life and its relationship with human life). Yet it is now over a decade since the initial promises of the HGP have been disturbed by the realization that the relationships between gene presence, genetic expression and function, protein synthesis, and both intracellular and extra-cellular environmental factors, are far more complex and indeterminate than initially imagined (Cook-Degan 1994; Kay 2000; Keller 2000; Rose 2007). The fact that the human genome possessed fewer genes than first imagined was proof enough that the relationship between genotype and phenotype was far more complex than simple one-way coding and instead, genetic function must work as part of a network of mutual interactions regulating gene behaviour and protein production; in short, gene ecology (Keller 2000; McNally and Glasner 2007; Lewontin 2000; Oyama 2000; Jablonka and Lamb 2005). As Keller suggests, one of the greatest achievements of revealing the codified map of human existence was not the ability of the map to disclose insights about genetic function, but instead the realization that humanity did not yet have access to an understanding of life's life-making processes – it revealed our paucity of knowledge rather than underscoring – through the book of life our wealth of knowledge. However this discovery did not reduce the broader rhetorical and symbolic power of the myth of genetic determinism, nor fantasies about its promise. Keller makes clear, in this sense, the routine coexistence of simplicity and complexity in the life sciences.

Wynne takes this realization a step further in his analysis of post-genomic biological knowledge production (Wynne 2005), by highlighting the existence of a further layer of reductionism often hidden by narratives of complexity in the life sciences. Despite an alertness to the complexity of post-genomic biology, he argues that continued and simultaneous reliance upon forms of reductionism may be fuelled not by stories of technical success, as in the case of CO1, but by the instrumental and commercial framings of much genomic research; 'The epistemic culture of instrumental technicism could, it seems, encompass and thereby selectively reduce any object of complexity, no matter how complex' (2005: 75). Wynne characterizes this instrumental, commodifying turn in the life sciences as an epistemic culture that requires analysis: he suggests that this culture of instrumental technicism 'does not encompass the questions about what is not deterministically predictable, or is uncontrolled; it excludes them' (ibid.: 77). In Wynne's rendition of the ways in which the complex and simple co-exist in post-genomic science, instrumentalism, commodification and control loom. We shall return to these issues as they affect barcoding in later chapters as we chart barcoding's key but, unlike Wynne's examples, not evidently commercial intersections with the 'ethically charged' (Rose 2007) territories of the global knowledge economy.

To connect this more general account with our narrative concerning the identification, or as we have suggested, *fabrication*, of CO1 as a standard global marker by which to identify all animals on earth, we reiterate what we have described so far. The naturalization of this gene and its remarkable affordances for the barcoding of animal species requires us to forget the process of assembling CO1, and to forget that CO1 is an achievement that could have been contestable at many points along the way. We have described above how part of CO1's success depended not only upon biological and technological labour but upon standardizing commitments and upon the vital enrolment of recruits to a taxonomic/democratic/salvational vision deriving from the taxonomic and broader funding communities. The ability for CO1 to do its job is not a natural feat but an achievement that is forged through the selective articulation of certain biological attributes, off-the-shelf primers and bio-informatic algorithms, the pre-existence of GenBank and much more taxonomic infrastructure, and a drive to translate the simplicity of the formula, 'one gene= one species = one name', into an elegant, economic, and visionary set of new global material linkages. The point here is that the birthing story of CO1 belies its artful alignment (Suchman 2000a and b, 2007) as a multiply situated object.

We are now ready to do the work of the second half of this chapter. Just as Hebert, in interview, revealed the decisions and choices that he had to make as precursors to the fabrication of CO1 as a universal genetic species-marker, we highlight below the contingencies encountered in making barcoding work for all life – not only for animals, but for plants as well. Dwelling on the search for the plant barcode, we describe the transition from simple to complex once again, from naturalized objects to denaturalized artefacts.

Genes behaving badly: the search for the plant barcode

Upon reading the first DNA barcoding publications reporting the success of CO1 for the species identification of animals, botanists quickly realized that a mitochondrial gene such as CO1 would not be appropriate for plants (Pennisi 2007). As botanists from the Royal Botanical Gardens at Kew stated, 'with plants we can't do it that way'.[21] Plants were different and, for them, for good biological reasons, CO1 was not going to work. This recognition immediately lay open to question the status of the iconic slogan, *one sequence = one barcode = all life*. With the projected extension of this formula to plants, the problem was partly that mitochondrial genes evolve more slowly in plants than in animals, and so do not display enough variation of DNA base-pair sequences between closely related species to distinguish them; yet this was the basic biological molecular marker for difference, and a species boundary. However, botanists' dissatisfaction with CO1 for plants rests upon a further set of biological and practical shortcomings of this gene, some of which we list in our Technical interlude and will not be fully elaborated upon here.[22]

When it became apparent that mitochondrial CO1 was not the appropriate marker with which to barcode plants, and that instead the plastid genome[23] was to be studied by botanists for more appropriate gene regions, the barcoding community was disappointed to say the least. If BOLI was really to fulfil its promise of barcoding all life, upon which so much hope had been staked, barcoders desperately needed the botanical community to be on board. Rather than the precise and universal simplicity that had been the barcoding of life story to date, the search for the so-called 'plant barcode' was proving to be a worryingly complex issue. A sense of urgency and growing tension flavoured the search for the plant barcode as it began to seem more elusive than previously imagined.

One aspect of the problem was that, for botanists, there had always been difficulties in getting plastid genes to yield signals of species boundaries. Researchers already knew that there was no 'magic region' in the plastid genome that evolves more rapidly, or rapidly enough to signal species differentiation. Therefore, there was less excitement for botanists around barcoding in general, because they had established reasons to be sceptical.[24] But second, and in direct contrast to this sense of fatalistic withdrawal coming from the botanists, a palpable pressure from the principal funders of BOLI – the Sloan Foundation – began to pervade the process of plant marker selection. In face of the emerging potentially terminal difficulties for finding a single plant DNA-marker akin to CO1 for animals, it was becoming clear that such vital financial and political support for BOLI as a whole was conditional on a decisive resolution. The emerging realization that CO1 would 'not work for plants' was of course the first hurdle, manifestly threatening the universalizing goals of barcoding. The governing infrastructure for BOLI, CBOL, therefore established a Plant Working Group (PWG)[25] supported by a grant provided specifically for this by the Sloan and Moore Foundations. In 2005, the PWG was tasked with selecting and testing a range of candidate loci for a faithful and universal genetic marker. Having agreed upon a selection of candidate loci, members of the PWG – from Europe, South America, South Africa and the USA – were given a number of months to experiment, with a view to arriving at a clear and evidence-based decision. Parallel to these efforts in what became known as the Sloan/Moore group, a team of botanists from the Smithsonian Institution simultaneously began testing and advocating different (and competing) genetic markers (Kress *et al.* 2005; Kress and Erickson 2007, 2008). The search became something of a race,[26] involving numerous candidate regions (and corresponding communities of botanists) vying for acceptance and visibility. But in contrast to the speed, relative isolation and singular focus with which Hebert and colleagues had made their CO1 'discoveries' on animal taxa in laboratories in Guelph, Canada, the labour and intellectual power required to come up with the solution for plants was distributed over a period of several years (officially between 2005 and the present), involving at least nine institutions, all over the globe, and testing 12 genetic regions (see Figure 3.1). Barcoding as a process of timely and efficient standardization looked like it was beginning to slow down and fracture.

It was not until 2008 that the CBOL Plant Working Group made public their decision on the plant barcode. On 7 October 2008, CBOL issued a 'progress to date' document on their website. This explained that on 25–26 September 2008, a meeting had taken place to decide once and for all upon the plant barcode: a decision had finally been made. Following official endorsement of the gene region by CBOL, and a publication by a selection of the botanists involved in the journal *Science*, the botanical community would now be able to deploy the plant barcode. The announcement made the selection of the plant barcode look easy, but in fact it belied three to four years of intense discussion and debate. We next recount the main negotiations preceding this decision.

The main players and principal events

Three most important properties were seen as required criteria for the selection of a plant DNA barcode region. These were stipulated as follows: significant species-level genetic variability and divergence; an appropriately short sequence length so as to facilitate DNA extraction and amplification and rapid sequencing; and the presence of conserved flanking sites for developing universal primers (Kress *et al.* 2005: 8370). However, early in the period of testing, data gathering and resulting discussions and negotiations, it quickly became apparent to all involved that no single locus (genetic region) adhered sufficiently to all three criteria. By mid-2008, seven candidate regions (trnH-psbA, rbcL, rpoC1, rpoB, matK, atpF-H, psbK-I) were still being seriously considered by several different research groups within the global botanical community. Figure 3.1 indicates, via the timeline on the left of the diagram, how enthusiasm for certain candidate loci cooled off, whereas for others it was maintained up until 2008.

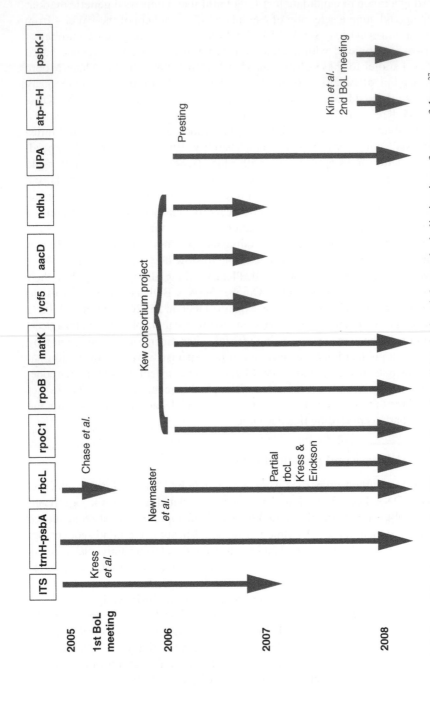

Figure 3.1 A graphic representation of the selection of candidate loci and the gradual elimination of some of these[27]

But even the gene regions that maintained interest until 2008 were found to be inadequate to work alone as the plant barcode. No single gene region from amongst these candidates in the plant kingdom was found to be able to perform the gargantuan task of representing all plants, as CO1 had managed to do for higher animals. What gradually became clear was that certain loci perform well according to certain criteria; other loci perform well to others. It was soon realized that the quest for a *single* plant barcode was the stuff of fantasy, suggested to be akin to the search for the Holy Grail (Rubinoff *et al.* 2006). As a result, a solution was proposed which would potentially disrupt and redefine the very concept of the DNA barcode. The plant barcode, as had already been surmised by a number of botanists,[28] would have to be *composite*, that is, made up of multiple gene loci. While entirely reasonable and 'evidence-based' in bare technical terms, this proposition would potentially introduce infrastructural mayhem to barcoding. It directly implied de-standardization, and a requirement for less controlled, distributed autonomous skilled knowledge and judgement, across networks of barcode-of-life users and developers. The potential unravelling of what was the intricate orchestration of a unitary, naturalized and universal world of biodiversity knowledge (and caring) seemed to be imminent.

In September 2007, an important and urgently planned PWG meeting was held during BOLI's second International Barcode of Life Conference in Taipei. Here the Sloan/Moore funded scientists, and the Kress team's proposals for a composite barcode were discussed within the barcoding community as a whole and a decision finalized as to the favoured gene loci to be used. The PWG meeting was a fascinating event to observe. As Figure 3.1 shows, five research groups had been involved in testing the different gene loci as potential plant barcode candidates. In a devoted plenary session, each team presented their own evidence and debated the advantages and disadvantages of their selected regions and their possible combinations in composite form. It was immediately apparent that there was no 'natural' front-runner: all regions had specific attributes; all created the possibility of detecting genetic variation that would map more-or-less faithfully onto known species boundaries. A final decision could not be made, however, as two crucial issues arose. Firstly, some of the regions proposed, although demonstrating faithful adherence to the biological criteria for selection, still proved extremely difficult to retrieve and align efficiently. After lengthy discussion, it was decided that this issue needed to be 'handed over' to a dedicated group of bioinformaticians (Data Analysis Working Group (DAWG)).[29]

As Kress and Erickson had foreseen, the real hurdle was the creation of algorithms for combining barcoding sequences from two or more DNA regions to yield species-level identifiers (Kress and Erickson 2007). The DAWG 'think tank' was to be specifically tasked with engaging as thoroughly as possible with plant biology and the idiosyncrasies of the preferred gene regions, so to be able to convert 'off-the-shelf' bioinformatic tools (algorithms) for DNA sequencing into bespoke gene region-specific devices.[30] The results of the DAWG study, it was hoped, would shed light upon which plant gene loci could realistically be sequenced, aligned and trusted to yield species identifications with precision and accuracy. What counted as 'realistic' and 'trustworthy' here, as technical matters, was conditional upon contingent assumptions about the imagined and unknown future barcoders' technical facilities, skills and disciplines, and their projects.

As we followed the progression of these debates, attending some of the DAWG meetings and discussions, we witnessed first-hand the ascendant role of bioinformatics in the decision-making process concerning the selection of the plant barcode, a role it had not played in the selection of CO1. The partial deferral by botanists to bioinformatic expertise highlighted, for example, the role of algorithms in achieving levels of confidence in species boundaries. The more important role for bioinformatics therefore was to foreground some of the ontological problems that have troubled barcoding from the outset (i.e. what is a species?), but which have tended to be subsumed by more practical questions and a shift of focus to DNA barcoding as a 'simple tool'. The relationship between the bioinformaticians and the botanists was not one of easy consent, however, as some botanists were dismayed by the computational 'reduction' of biological complexity; one botanist stated for example that algorithmically mediated approaches to species discrimination on the basis of sequence divergence amounted to '*no more than a probability statement*'.[31] Here the ability of the possible plant barcode to do its job required a traversing of increasingly long chains of inferences and imputations associated with DNA sequencing algorithms and the carefully calibrated negotiations between botanists and bioinformaticians, the details of which we have not been able to do justice to here. Unfortunately for the PWG, DAWG discussions on the selection of a plant barcode would need to continue in order to reach a final agreement satisfactory to bioinformaticians and botanists alike.

Returning to the general process of decision-making, the fact that a decision regarding these loci could not be made was one setback for the researchers present at the PWG meeting at Taipei in September 2007. A second 'hiccup' that stalled the day's proceedings came from South Korean researchers hitherto unknown to the rest of the PWG. Towards the end of the PWG meeting Dr Kim asked for the microphone. He then stood up and presented his team's research results.[32] These quite unexpectedly suggested the need to consider a combination of gene regions that had not been seriously considered by either the Sloan/Moore or Kress research teams. Whilst the introduction of this possibility came as a surprise, and would further slowdown the process of reaching a final decision, it could not be ignored. Dr Kim and his team had yet to test their findings on a sufficient set of land plant samples. In Taipei, the PWG therefore decided that a final decision on the plant barcode would be suspended pending the publication of Dr Kim's results, plus the results from further work to be undertaken by DAWG. The date and place provisionally set for this was San Diego, June 2008 when the Scientific Advisory board and DAWG would convene to discuss this and other barcoding related issues.

Two issues stand out from the Taipei conference. First, barcoders were learning to engage in co-operative craft-work and to 'air' this craft work – surprises, contingencies and all – in plenary meetings at BOLI conferences. Second, such collaborative work and the contingent unknowns and surprises that were met along the way, meant that barcoding was slowing down. Standardization often involves extensive unseen technical and political work, including projection of imaginaries of diverse future users and their situations (Bowker 2005). Furthermore, as Thévenot notes, standards provide 'a guarantee' by attributing properties to standardized objects, even while such guarantees rest on an unsure dependency between agent and environment (Thévenot 2009: 807). The more ambitious the scope of such standardizing projects, the more extensive and complex is such work likely to be. As the plant story shows, standardizing the means of identifying life itself was no exception.

What does the search for the plant barcode tell us?

By laying bare a representative part of the negotiations of botanists in barcoding, we are able to show more clearly the delicate artifice needed to create the plant barcode. Botanists faced the challenge of 'artfully aligning' (Suchman 2000, 2007; Suchman and Bishop) the universalizing aims of barcoding, the aims and requirements of barcoding's funders, the properties of plant gene regions, the capacities of sequencing and alignment technologies, and the knowledge and skills of bioinformaticians and taxonomists. Appreciating this, and the collaborative work that it took, we can perhaps more easily 'see' the fabricated, artefactual and co-produced nature of CO1 – the gene that performed so well for animals that it spawned the dream to DNA barcode *all* life. In a sense, the plant story shows how the icon of the barcode itself was temporarily brought down to earth. CO1, as icon, was seen as a universalizing and singular, natural, galvanizing entity – an 'index' for the variability of all animal species on the planet. CO1, as we shall explore, was so successful that entire infrastructures, such as BOLD were initially designed around it (Chapter 6). But what the plant barcode story also shows is that CO1's heroic properties as an 'organizing principle' were the product of a fortuitous set of contingencies that slotted perfectly into place under the idea that twenty-first-century taxonomy must – rapidly! – perform to the new rubrics of speed, economy, universality, circulability, simplicity and *non-specialist* utility.

Since CO1 performed these demands so apparently effortlessly, the task at hand became one of promoting CO1 and the technoscientific infrastructures that supported it, as a global standard. What we have seen in this chapter, is that just as BOLI and CBOL were 'naturalizing' CO1 as 'the' standard for twenty-first-century DNA barcoding, those searching for the plant barcode were simultaneously breaking this barcoding standard down. As it faced the exigencies of plant DNA (and by association, plant evolution), barcoding as a whole had to become more heterogeneous and complex. In facing and responding to a breakdown in standard universalizing techniques, however, it perhaps became more robust, as a science and as a vision. Barcoding had to become less hasty to achieve its universalizing ambitions, and to get more involved, again, in research. We might suggest here that universality does not have to mean singularity – that coordination across difference is imaginable, requiring as it does, distributed skills, appropriate commitments, and a commensurate moral imagination. In the process of development to encompass the plant kingdom, barcoding had to open up single laboratory 'discoveries' to new collaborative research partnerships that brought the know-how and craft of different disciplines (e.g. in the making of algorithms) into the science.

It is also clear that, in this process, the quest for the barcode was more genera-tive than anticipated. A whole set of new questions concerning the role/use of genes in identifying species, generated by the intensifying relationship between biologists, bioinformaticians and data managers, was unleashed in a way which resonates with Evelyn Fox Keller's descriptions of the de-coding of the human genome (Keller 2000; McNally and Glasner 2007). Keller notes that this gen-erativity – the spawning of new questions and openings, and new collaborative networks and programmes for science – strangely did not detract from the con-tinuing power of the characterization of the human genome project as a 'one gene = one function', deterministic endeavour.[33] In many ways this is also true of barcoding: Hebert's strong narrative of 'one gene = all species = all life' somehow survived the search for the plant barcode. It has become a more open but unify-ing *heuristic* (as distinct from a literal claim), that collectivises and facilitates the disparate particular agendas, ontologies, and epistemic cultures comprising the BOLI project. Thus, the ontology of CO1 became highly complex: it was both naturalized (through these strong narratives) and de-naturalized (through the open and difficult search for the plant barcode), at one and the same time. Here we can also glimpse the simultaneous technoscientific performance of reduction and complexification, discussed earlier. But barcoding also had to slow down in this process of 'ontological choreography' (Thompson 2005). It had to acknowl-edge legitimate difference, even unruliness, in nature. It had to acknowledge these qualities also within the taxonomic communities, and to make transparent the emerging research findings, and choices that could be made. As we have sug-gested, these particular difficulties relating to the search for the plant barcode arguably strengthened rather than weakened the phenomenon of DNA barcoding. By laying bare the human-technological – biological relationships that needed to be nurtured in order to have the plant barcode up and running, the status of barcoding per se – and more importantly its surrounding human and technologi-cal infrastructure – were made more robust, less brittle, and more inclusive. A lack of standardization in genes themselves turned from appearing to be a terrible weakness to becoming a particular strength (the apparent 'unruliness' of plant genes can perhaps be usefully perceived as an example of biology 'kicking back' (Barad 2003, 2007)). Barcoding had to go back to the labs, and as a result, it had to mature as a science. Ironically perhaps DNA barcoding would have lacked this opportunity had it not been for the (plant) genes that behaved so badly.

Finally, although it was of little interest to the taxonomists working to resolve these issues, the entire debate about the plant barcode exposed quite beautifully the collaborative crafting and skilful artifice required to find both animal and plant genetic regions and to fit them with bespoke algorithmic tools, in order that they might yield faithful correspondence to already established species identifi-cations from morphological and related practices. This was fascinating to us as observers, giving us a real insight into the way in which barcoding is retro-fitted to established classes and nomenclatures. Barcoding, at this level and for this particular quest, does indeed work hard to create the controversial 'gene-species' (half DNA/half morphological taxonomy, mapping genetic barcodes onto established Linnaean names) that concerned Vogler and Monahan, as we described in Chapter 2. The quest for the plant barcode, within BOLI and CBOL, avowedly looked no further than this identification end. As Hollingsworth *et al.* (2005) suggest, achieving this by using multiple markers was the 'crux challenge' for the botanical community in the mid 2000s. Botanists like Kress had a crystal clear idea of the purpose and value of barcoding: '[T]he main purpose of DNA barcoding is not to build phylogenetic trees, but to provide rapid and accurate identifications of unidentified organisms whose DNA barcodes have already been registered in a sequence library' (Kress and Erikson 2007: 508). Thus, the search for the plant barcode shows us that, not only was CO1 the subject of repeated social acts of naturalization, such that its fabrication was deleted and its natural efficacy came at times to be utterly taken for granted. Likewise, barcoding itself, informed by a highly limited version of a taxonomy 'fit for the twenty-first century', came to earn an unquestioned status as a practical, standardized tool that would be put to use in biodiversity assessments, ecological studies and forensic analyses, rather than in animal or plant taxonomy per se (Kress *et al.* 2005). Paradoxically, CO1 as heroic icon 'embedded in every cell' survived the search for the plant barcode, only to support these most modest twenty-first-century user-goals. In subsequent chapters we analyse this blending of hype and modesty so characteristic of the genomic sciences.

4 'A leg away for DNA'

Mobilizing, compiling and purifying material for DNA barcoding

An inspector at a busy seaport, a hiker on a mountain trail, or a scientist in a lab could insert a sample containing DNA – a snippet of whisker, say, or the leg of an insect – into the device, which would detect the sequence of nucleic acids in the barcode segment. This information would be relayed instantly to a reference database, a public library of DNA barcodes, which would respond with the specimen's name, photograph and description. Anyone, anywhere, could identify species and could also learn whether some living thing belongs to a species no one has ever recognized before.

(Stoekle and Hebert 2008: 83)

The vision of a ubiquitous, freely accessible and pocket-sized technology – the 'barcorder' which instantaneously converts snippets of nature into verified species names – imagined in Stoekle and Hebert's words above – is a powerful and familiar one in DNA barcoding circles. There is more than a vision condensed in these words however. This portrayal of a plucking of organisms and the speedy retrieval of their correct identification understates the enormity and complexity of a globally distributed, disciplined network of *collaboration* which this vision requires; and it does not speak at all of the many and varied social, organizational, geographical, logistic, technically uncertain and politically sensitive *modes of access* to biological material required to feed and nurture the science and infrastructure of BOLI. In Chapter 3 we focused upon ways in which distributed collaborations between taxonomists, lab managers and bioinformaticians have been reconfigured by BOLI in accordance with barcoding's need for standardizing techniques and technologies. In this chapter we consider a range of collaborations required for BOLI's procurement of materials for the digital inventorying and archiving of life as barcodes. We thus bring to the fore the materiality and material practices of the venture, as formative of and not only secondary to the scientific knowledge dimensions (Knorr Cetina, 1999). We simultaneously focus upon ways in which BOLI is required to work with and through the *socio-economic and political histories* within which the collection of materials have been – and still are – tightly entangled.

A brief glance at BOLD reveals to the untrained eye a bewildering array of barcoding projects each of which generates a number of fields including 'taxonomy', 'geography', 'attribution fields', specimen depositories, dataset codes, specimen and sequence identifiers, and so on. All of these fields are vital for ensuring that DNA barcodes are traceable back to the organisms represented by the barcode (Ivanova *et al.* 2006; Ratnasingham and Hebert 2007). At the time of writing, 8,200 projects are currently archived in BOLD, amassing a total of 1,733,466 Barcode Sequences representing 116,917 animal, 40,507 plant and 2,386 fungi and 'other life' species.[1] A fuller account of the careful compilation and uses of BOLD is provided in Chapter 6. But we emphasize here the way in which BOLD indicates the quite extraordinarily variable array of sites from which biological materials destined for barcoding have been derived. Also and perhaps in a format more visible than BOLD, the Canadian Centre for DNA barcoding (CCDB) credits a range of partner institutions (museums, universities, policy agencies) worldwide for access to their specimen collections. A few of these are visible in Figure 4.1.

BOLI may in some ways be a universalizing initiative, as we emphasize in Chapter 3, but its sites of production are highly differentiated and distributed, so that the multiple forms of standardization which BOLI organizes across such a network also have to accommodate to some degree the differences which grounding involves. Thus, standardization paradoxically always depends on an accommodation of, and with, diversity (Waterton and Wynne 1996), as our examination of BOLI shows. These institutions and locales, sites of natural history knowledge-making, underline the importance of the local, the material and the immaterial (the symbolic, the negotiated) in the making of universal standards for BOLI.

Museums, herbaria, individual collections, wild nature and national parks are examples of the many locations from which specimens are sought, harvested, prepared and processed. But by simply noting the places of specimen-provenance it is easy to forget that this variable nature is also variably social. Obtaining access to material, the meticulous preparation of specimens in accordance with taxonomic (and barcoding standards), and the mobilization of these materials through multiple transfers and translations from 'the field' to the robotic sequencing laboratories, is not only hard work but involves careful negotiation with collection curators, laboratory technicians and managers, field biologists and sometimes, as we shall explore below, indigenous people inhabiting the most bio-diverse areas of the world. All of these actors and more are considered and implicated in different ways in the production of DNA barcodes.

Observing the dispersed provenance of BOLI's materials in this way allows us to appreciate the creative labour invested by BOLI and its partners in building collaborations – enrolling the 'rest of the world' to supply the raw material required to enable and sustain it as a viable genomics technoscience. It also allows us to perceive the distribution of BOLI's sites of innovation and the ways in which its activities are markedly de-centred. We are attentive to how BOLI extends its networks as *distributed sites of innovation* and how, in so doing, it finds itself straddling the locations and temporalities of fast, hyper-genomic, big science

Figure 4.1 CCDB partners http://www.ccdb.ca/pa/ge/about-us/partnerships/partner-speci
men-collections

Source: Permission to reproduce image granted by Sujeevan Ratnasingham, CCDB,
Ontario Biodiversity Institute, University of Guelph.

and the different worlds of local scientific practice, local community livelihoods,
international treaties, and global biodiversity politics and policy. A focus on the
proliferation of BOLI's sites of production provides a more accurate depiction of
the topographies of innovation and of an increasingly global division of labour
(Ong and Collier 2005; Suchman 2002). We explore below how its novelty as an
innovation is not only framed by its geo-spatial distribution. It is also framed by
the deep inheritance of centuries of collection practices. We show how BOLI both
draws upon, but also diverges from, or 'forgets', the histories of global regimes of
collection, production and ownership (Grove 1995; Hayden 2003a; Richards 1993;
Thomas 1984) as it seeks integration between multiple sites, multiple materials and
multiple human players, in pursuit of a singular technoscience-centred vision.

Parry (2004) documents how imperial and economic expansion has, since the
seventeenth century, conjoined imperial progress with the collection of biological
and geological resources from the global 'South'. She traces how, since the 1980s
and 1990s, historical modes of resource-extraction shifted towards a 'new nature
hunt', *bio-prospecting* (ibid.: 13), as resources were routinely transferred in the

same geo-political direction but for more explicitly commercial biotechnological processes. These movements of material, she states; 'not only mirror earlier colonial projects of appropriation, but in fact constitute a new and even more pervasive form of "bio-colonialism" or "bio-imperialism"' (ibid.). In 1992, the signing of the UN Convention on Biological Diversity (CBD) at the Rio Earth Summit by 193 nations was a clear response to a need to transparently regulate access to biological resources, a need accentuated by the post-colonial politicization of nation states and local communities sensitive to the power-asymmetries and exploitative histories of colonial regimes of resource-extraction.[2] As Parry (2004), Hayden (2003a) and Greene (2004) observe, bio-prospecting companies have however found myriad ways of evading these regulations both through naivety and by deliberate denial.

As we further discuss in Chapter 5, while Parry was referring to the central goal of this neo-colonial bio-front as a commercial one, BOLI has professed no such purposes, and indeed as we describe below works hard to distinguish itself from such ventures, as a non-commercial, protection-oriented science doing good for source-countries and communities (Schindel 2010). We (and BOLI) are however alert to Parry's observations that the ends to which collections are put can rapidly shift in purpose, and the Nagoya Protocol for Access and Benefit Sharing (ABS) (CBD 2010) has been designed to regulate such potential shifts.[3] Although potential commercialization of barcoding is not our focus in this chapter, we do need to describe how BOLI has understood, and navigated the troubled histories of collection. Pivotal to these histories are those materials in museums of natural history which are witnesses to past relations of collecting which would now be impermissible, and which are in some cases under disputed rights of ownership as countries of origin pursue repatriation claims. Those vast collections however, are rich research resources for BOLI, and as such are subject to contemporary Material Transfer Agreements. As we observe BOLI's collaborations with indigenous communities, museum collections curators, and the access and benefit sharing aspects of the CBD itself, we ask how its approaches to collecting demonstrate continuities with and reconfigurations of the troubled biopolitical histories and relationships mentioned here. We explore the ways in which BOLI as a unifying, standardizing global project engages explicitly with some of the intricacies of such 'local' political complexity, whilst perhaps not recognizing, or even denying the saliency of others. Tsing's (2005) ethnographic account of the 'frictions' of local-global collaborations is relevant here. She explores how the negotiated politics of collaborations themselves may often be 'forgotten' once their products attain a certain visibility and credibility:

> Buoyed by axioms of unity, collaborations create convincingly agreed upon observations and facts that then appear to support generalizations directly, that is, without the prior mediation of the collaboration. The contingency of the collaboration, and its exclusions, no longer seem relevant because the facts come to 'speak for themselves'.
>
> (Tsing 2005: 91)

We ask in this chapter if BOLI, with its need and 'thirst for mixed connections' (Latour 2010: 481), pays due attention to the biopolitical arrangements it aids and abets in proliferating. For later discussion, we also ask to what extent the contradictions which we identify in BOLI's practices of collection, are in fact constitutive elements of such a self-confident 'public good' technoscience? And, if so, how might BOLI itself address such contradictions? We begin our analysis of three collaborations, selected from a large range of possibilities, by looking at the engagement by BOLI of Costa Rican indigenous communities as producers and providers of biological specimens.

Parataxonomist social labour for BOLI

It was when we were in the office of an eminent entomologist and key BOLI player at the Smithsonian in Washington that we became aware of the reliance of BOLI upon many individuals and institutions working diligently the world over to amass, prepare and sequence specimens required to feed and nurture BOLI. Whilst watching the entomologist's assistant bent over her microscope carefully removing butterfly and moth legs destined for DNA sequencing, we noticed a tower of solid, somewhat forbidding plastic crates, fastened at the top with red and grey heavy clasps part-blocking the way out of the room. These rather unwieldy and cumbersome boxes somehow occupied an unexpected space amongst the more familiar arrangements in the office. Happily for the entomologist and his assistant they had just arrived; the latest shipment of specimens from Daniel Janzen in Costa Rica.

We were aware from the very start of our research of the fact that Daniel Janzen had played a pivotal role, in both visionary and material ways, in propelling BOLI. As Holloway (2006) narrates, Janzen and Paul Hebert met in 2003 at the Banbury Cold Spring Harbor Laboratory when Hebert was first introducing the new concept of DNA barcoding to the taxonomic community. Janzen ('with 16 butterflies hidden in his briefcase')[4] spoke from the floor requesting a rapid, cheap and high-resolution identification tool. This was to be no bigger than the comb he removed from his pocket in a flourish of demonstration, and would immeasurably help his work in undertaking a biodiversity inventory of Costa Rica's Area de Conservaciòn de Guanacaste (ACG). With comb held aloft to make his point, Janzen was referring to the idea of the hand-held barcorder, a mobile sequencing and computing technology yet-to-be-invented, but much vaunted, and referred to many times during the course of our research as (imagined) technology that is 'not far off'. The sheer enthusiasm that Janzen had for archiving his ecological findings amongst the bio-diverse 'hotspots' of Costa Rica was exactly what Hebert needed as a testing ground for the efficacy of his CO1 gene-segment marker, and he agreed, as a starter, to sequence a selection of Janzen's specimens of the neotropical skipper butterfly – *Astraptes fulgerator*. This species had been perplexing Janzen for some time as he and his colleagues were aware of differences in caterpillar colour patterns, too divergent for a single species, which they had begun to match with a variety of food plants. However, such differences were not reflected in the male or female genitalia usually dissected as clues to cryptic differentiation. Hebert and Janzen were both excited by the results of the use of the CO1 barcode as a diagnostic tool, as it revealed that Janzen's *A. Fulgerator* in fact included 10 distinct species. The publication resulting from this research 'Ten Species in One: DNA barcdoding reveals cryptic species in the neotropical skipper butterfly *Astraptes fulgerator*' (Hebert *et al.* 2004a), signalled the potential of BOLI and sparked the hope that many more hidden (or 'cryptic') species might be waiting to be deciphered and revealed – a possibility which captured the public, funding and scientific imagination.[5]

Returning now to our observations at the Smithsonian, the entomologist's assistant explained to us that inside the large boxes were neatly stacked caterpillars and butterflies collected and arranged not by professional taxonomists but by local, indigenous 'parataxonomists' who had been trained up by the project in location in several Costa Rican villages.[6] According to her, they had been reared, observed, anaesthetized, killed and pinned by 'families living in the forest'. Indigenous people were not being tapped here for their local knowledge – a practice that is becoming conventional in many global-local biodiversity projects (Hayden 2003a; Greene 2004). Instead, as part of the task of undertaking a biodiversity inventory in areas of complex and under recorded natural diversity, whole families were (and are) being trained in field collecting techniques and in data management methods. Indeed the 'Ten Species in One' article refers, albeit rather fleetingly, to 'the rearing of >2,500 wild-caught caterpillars' upon which the research depended. The short reference suggests that as more cryptic life is revealed by DNA barcoding, it is in fact *bred* to expose further differentiation; materials are not only 'collected from nature' in the conventional sense of the term, they are in fact being produced for BOLI.

The realization that local communities were being actively enrolled by BOLI as *producers of organisms* was a surprise to us. It raises the issue of how Janzen and BOLI might be reconfiguring a historically established global distribution of labour designed to facilitate access to biological material. Such access is currently mediated by the Nagoya Protocol of CBD and in the case of the ACG, Janzen and colleagues had adhered to requirements established by the then-current CBD legislation for gaining legitimate access to biological materials in areas inhabited by local communities. What interests us here however are not issues of access to material (and financial recognition of local or national provision)[7] per se; we focus instead on what we perceive to be the *social labour* required (and imagined by BOLI scientists) to provide material for BOLI – a type of labour which, we suggest, exceeds the remit and imagination of the CBD framing of value and contribution. By social labour we refer here to ways in which local individuals and communities are seen to be investing and changing their own *social relationships and their very selves* through the process of rearing specimens for barcoding research, and ultimately for global biodiversity conservation. This focus allows us to flesh out – in ways we had not anticipated – the very particular manner in which BOLI was managing to construct an infrastructure of distributed and disciplined work, towards its own global technoscientific and further goals.

Leaving the Smithsonian eager to know more about the parataxonomists, we came across two publications, both revealing the importance of local people in Costa Rica and Papua New Guinea as producers and providers of material for BOLI (Bassett *et al.* 2004; Janzen 2004a). Through these, we observe here the ways in which Janzen, as one example of a scientific practitioner in the field who has taken up barcoding with enthusiasm, evokes the intricacies and idiosyncrasies of what is involved *and imagined* by BOLI scientists in establishing this very particular type of collaboration for the accumulation of specimens.

Janzen does three things in his short but powerful article. First, he celebrates the ingenuity of enrolling local people to expand the labour force required to under-take what is to be a mammoth inventory of Costa Rica's life-forms. Immediately, but we suspect unwittingly, establishing a neo-colonial tone, he acknowledges the fact that collectors visiting mega-diverse regions of the world have of course long depended on local residents for labour, information and material samples. After all, he asks, rhetorically, 'Who was to carry out this very labour-intensive task in a hot, wet, unfriendly tropical rural environment?' (2004a: 182). He reminds the reader of the serious practical challenges of training local people in collecting, rearing and preparing specimens, and of furnishing them with a series of compe-tences required to meet global taxonomic standards. Second, he writes about the parallel need to recognize the moral, socio-economic realities of employed local individuals and their families. In the following quotation, Janzen expresses how parataxonomists are a community that is embedded in the close kinship world of their extended families and at the same time are caught up in the whirl of the salvationary aspirations of the CBD:

> the parataxonomist may well be marching to an international drumbeat encoded in the Convention of Biological Diversity and centuries of academic biological query, but he or she has to survive largely in the social milieu of a village resident on the margin of the fearful wilderness called a national park or other kind of conservation area.
>
> (2004a: 184)

Such a liminal position, Janzen reflects, calls for sensitive recognition of the fact that parataxonomists have to sustain a livelihood, manage an immediate and familiar social environment itself coloured by history, and lastly (and most excit-ingly for Janzen), extend their gaze and responsibilities outwards towards centres of taxonomic research and accumulation such as the Smithsonian Institution or the American Museum of Natural History. From Janzen's perspective it seems that the collection of biological specimens meets only half of BOLI's and indeed the CBD's real requirements; global biodiversity needs a change across the board and, as we suggest above, this includes a consideration of the *social* (and norma-tive) investment involved. Or as Janzen puts it, 'vertical contracts' need to be established here in the sweaty Costa Rican rainforest between local families and the 'taxasphere':

The first act of raw conservation, critical as it is, in no way guarantees conservation into perpetuity unless the second act is integration of that wildland into the social and economic fabric of the resident, national and international community.

(ibid.: 186)

Janzen continues, and in the same paragraph, details the kinship, marriage and residence patterns of one particular community and the specific rearing and feeding requirements of the five caterpillar barns under their care and responsibility. But he also, somewhat wistfully, comments on the affective shift he believes is experienced by partaxonomists as they commit to a world of species identification:

For example, the thrill of finding a new species, for the parataxonomist, comes largely through observing and building on the thrill expressed by his or her new employers (and the attendant trickle down through that line of security), rather than in the form of enthusiasm by parents or neighbors who first could not care less (a new species does not put bread on the table for them) and second, have probably been quashing that new species underfoot for generations.

(ibid.: 184)

This last quotation introduces in bare terms Janzen's third aim – to reconfigure and fold in to the BOLI collaboration, distinctly preferred ways of being human. We look further at this subtle but unsettling dimension of Janzen's (and by implication BOLI's) commitment to barcoding as a means for saving global biodiversity. In his ambitions to fulfill the latter long-term goal, Janzen unashamedly, and for him necessarily, devalues local livelihood strategies. The work for the 'taxasphere' not only pays better than washing clothes and subsistence agriculture. It introduces a moral imperative to abandon not only detrimental ways of valuing and exploiting nature but also to abandon that social-cultural milieu. The social lives and livelihoods of the Costa Rican caterpillar rearers ('gusaneros') are simultaneously recognized by Janzen to constitute the source of biodiversity extermination, largely through lack of care.[8] In other writings (Janzen 2004b) Janzen extends a similar accusation to most of humankind, connecting a lack of care to a lack of knowing or 'bio-illiteracy'. He writes of a destructive, bio-illiterate, global public that he believes could be converted to a loving 'biophilic' relationship with nature (as first propounded by E.O. Wilson: Wilson 1984, 1991) precisely through an increase in their knowledge of species, or their 'bioliteracy', through barcoding (Ellis *et al.* 2010).

It seems quite clear that the Costa Rican parataxonomists are literally producing biodiversity for BOLI, and thus present evidence of the continuation of a historical pattern of extraction of raw materials, human knowledge and labour (Hayden 2003a; McNeil 2005; Parry 2004). *Biodiversity has to be bred in order to be read.* But there is something else revealed by Janzen's reflections about what it has meant for nature and humanity to enroll Costa Rican 'gusaneros' in the task of rearing caterpillars, killing them and their adult variations to feed the BOLI appetite for samples, so as to expand human knowledge of and care for biodiversity. Janzen is explicit in interweaving the merits of bioliteracy for human cognition and morality, with livelihood strategies and a shift towards hi-tech dependent living. The direct engagement of 'gusaneros' is part of a (philanthropically intended) move to convert laundry-pounding, insectcrushing, bean-picking inelegance (and implicit backwardness) to the elegance of biodiversity-knowing and biodiversity-valuing beings. This normative force is invested in the technoscience of BOLI.

By gliding between the epistemic, material and socio-economic needs of scientists and parataxonomists in an unnervingly seamless manner, Janzen describes a future biodiversity-conservation-to-come which depends upon a radical process of conversion, requiring different ways of being human and nurturing very different types of relationship between humans and the non-human world. Interestingly, this does not so much concern questions about the alienation of parataxonomists from the products of their labour – a key concern of the CBD, especially since the Nagoya protocol. Rather, his portrayal of the ACG parataxonomists is about an estrangement – in an imagined and constructed sense – from the whole sense of self and relationships to kin and nature through a process of conversion. Furthermore, it is also apparent that the organization of labour and conversion of Costa Rican parataxonomists currently allows little room for appreciating and drawing from a range of existing, unacknowledged and untapped human sensibilities developed *through* relationships with kin, livelihoods and biodiversity itself. Correct and alternative ways of being human are overtly prescribed by BOLI, an observation which raises the question as to whether BOLI could possibly undertake the inverse move – to subscribe to understand existing social relationships, including social relationships with nature, and to explore how to build in quite different ways upon these more culturally grounded, more diverse and differentiated yet possibly more robust, foundations (Haraway 2007, 2010; Latour 2011). As we describe in Chapter 9 we see this kind of move as requiring a different kind of normative or moral imagination to that currently evident in BOLI, which would also potentially develop new social relations and practices as working parts of BOLI, with implications also for the development of BOLI as technoscience.

We have drawn at length here on Janzen's writing and have been critical of the way in which it (and by inference, BOLI) incorporates a particular vision of *human subjectivity* and *social labour* into the demanding work of *building collaboration* as part of a distributed technoscientific program. On the one hand Janzen is very conscious in his article of the social milieu within which biodiversity research is taking place in the ACG. On the other, in so doing, he is completely unreflective about the neo-colonial framing of his own assumptions and the violence which this does to past histories of subjugation and abuse of indigenous peoples as part of regimes of specimen collecting – whether these have been in the name of empire building or of biodiversity conservation. What we wish to emphasize throughout this chapter and through this first example, is in part the ingenuity of BOLI scientists in the ways in which they distribute and tend their sites of innovation. But what we also illustrate here is that the flow of genetic material to BOLI which is its foremost need, is complicated by intricate particular social relationships which are imagined by BOLI scientists to be required. We are not arguing that considerations of such relationships are not warranted; quite the reverse. We suggest that, in the extension of its innovatory networks, BOLI needs to do more in this regard. It needs to interrogate its own assumptions and practices as it retreads the paths of past natural history collectors which by now, it is widely acknowledged, did much more than accumulate the wonders of the natural world in centres, museums, botanical gardens and sanctuaries far removed from the biodiversity rich regions that fed them. They also used them in ways that often delivered little or no reciprocal benefit. In the case narrated, of Janzen and his butterflies, the engaged parataxonomists have, to our knowledge, posed little-or-no resistance to this process, and BOLI has been able to forge ahead down this particular route of specimen collection. Our second case changes direction as we consider BOLI's harvesting, not of fresh specimens from the mediated 'wild' but the equally important (re)collection of specimens archived for centuries in the world's herbaria and natural history museums.

(Re)sampling museum specimens and the All Birds Barcoding Initiative

From the list of sites of provenance of BOLI's materials referred to above (Figure 4.1), the importance of museum collections for compiling a universal archive of DNA barcodes is obvious. The world's many regional and local natural history museums, amplified by the huge imperial collections in developed world centres such as London, Paris, New York and Berlin, are seen as treasure-troves of collected and archived life forms. These may have open doors for public education and wonderment, but remain in their material collections of life largely unavailable to anyone except museum specialists. It is perhaps no coincidence that the impetus driving BOLI – via DNA – to digitally archive all species on the planet – corresponds with the new commitments of many historic natural history museums across the world to digitally document their treasures and to 'open up' their collections to a wider, potentially global, public via the world-wide web. Needless to say such a commitment implies a huge mobilization of labour, materials and resources (time and money) and involves natural history museums in having to re-think their roles and priorities not only as the 'treasure houses' of the natural world, but also as the places where many taxonomists are currently employed, as researchers and as curators of those collections.

In this section of the chapter we select a specific barcoding initiative – the All Birds Barcoding Initiative (ABBI) – through which we reflect upon the relationship between BOLI and archives of dried, classified and meticulously curated specimens. ABBI is an example of a global project launched in September 2005 with aims 'to collect standardized genetic data in the form of DNA barcodes from approximately 10,000 known species of world birds'.[9] The ABBI website notes that, in spite of the popularity and relative ease of bird taxonomy, hundreds of species remain undescribed. Within the USA and Canada however, most if not all species have already been collected, identified and arranged in various natural history museums, including the American Museum of Natural History (AMNH) and the Smithsonian's National Museum of Natural History (NMNH). A subsidiary project within the ABBI, was a project compiled between the NMNH (specifically Carla Dove, the Program Manager of the Division of Birds, see Figure 4.2) and BOLI. We mention this project partly because it involved an interesting collaboration between the US Federal Aviation Authority (FAA), the Smithsonian NMNH, BOLI and a number of North American museums housing and curating North American bird specimens (see Chapter 8 for further elaboration upon the multiple and future practical *uses* for DNA barcodes).

The main drive behind the project was the growing concern on the part of the FAA about the impact of 'bird strikes' (birds that have collided with aircraft) on avian safety. They realized that the rapid and accurate identification of mashed-up samples of bird-flesh and feather sent to Dove's laboratory would help in the design and deployment of adequate safety measures such as a re-routing of aircraft to avoid known 'culprit' bird communities.[10] This would be a job easily accomplished by DNA barcoding, allowing an immediate correlation between the unidentifiable remains scraped from an aircraft's engine and the DNA from a *known* bird specimen carefully dried and curated in a museum collection. For this chapter this ABBI venture is significant because the process required painstaking negotiations with the curators of bird collections across the USA and Canada in order to access the required material. We briefly reflect upon the concerns of some collection curators expressed at an ABBI workshop in 2005[11] and by one in particular during an interview with us in 2008.[12]

First, upon understanding the demands which the ABBI (and other cognate barcoding projects) would make upon curators and the often fragile contents of their collections, some individuals present at the workshop voiced surprise at BOLI's sense of *entitlement* to an easy and universal access to material. Although materials housed in public institutions are generally accessible for scientific research and this access is regulated by institution-specific Material Transfer Agreements, the enormous amount of material wanted and the rigor of practice introduced by BOLI's particular standards, placed new demands on the already heavy workloads of museum practitioners (as well as placing at risk the quality and endurance of the specimens themselves). It is worth noting here that this has been a more global concern – certain lab managers at other BOLI global outreach and Leading Labs events have also complained of BOLI's unfair expectations upon groups with their own priorities, and limited resources and time. BOLI's increasingly urgent requests for material and a speedy mobilization of this through laboratory 'pipelines' (DNA preparation, sequencing and the meticulous auditing of every stage in the process) had, by 2008, in the case of the Smithsonian's Laboratory for Analytical Biology (LAB), produced a 'bottleneck' of semi-processed material and information which was not only slowing down BOLI's progress but also LAB's parallel research.[13] These anxieties illustrate some 'friction' between

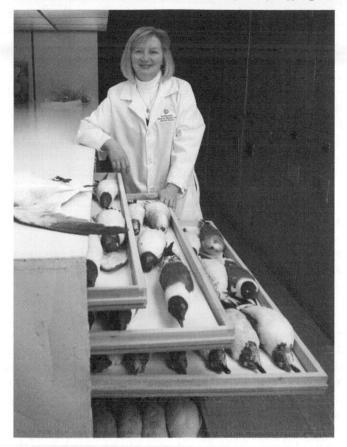

Figure 4.2 Dr Carla Dove, Program Manager of the Division of Birds, National Museum of Natural History, Smithsonian Institution

Source: We are grateful to Donald E. Hurlbert for permission to reproduce his photograph. Image courtesy of Carla Dove.

BOLI's ambitions for speed and streamlined collaboration and the material and human-technological relations of production these require. They also highlight a mis-matching appreciation of the value of historical and current day accumulation of human, biological and technological labour. Whilst museum curators are fully invested in an understanding of past and present labours needed to bring nature into an archive such as a database or even a museum collection, BOLI has not perhaps fully considered or folded in to their accounting and planning, the labour required to open up the (public) value of DNA barcodes in this way.

Second, curators wanted to know and be able to control exactly what the material they would be relinquishing for DNA barcoding would be used for, and how quickly data would enter the public domain. They were worried, for example, that genetic resources under their jurisdiction might be used by BOLI scientists for *more than* CO1 extraction for mere identification work. They feared that Material Transfer Agreements and the uses of material they stipulate in each case, would be evaded as BOLI scientists might undertake and publish more complex phylogenetic research using these samples. This takes us back to Parry's (2004) consideration of the politics of 'micro-sourcing' as well as to taxonomic debates concerning how limited or expansive the remit of DNA barcoding should be. Parry (2004), in her detailed observations of historical and contemporary museum collecting practices, describes collections as performing a 'concentration, disciplining, circulation and regulation of assemblages of material' (2004: 15). As such, she explores the ways in which museum and herbaria specimens – be these whole organism collections or cryogenically stored tissues samples – are decontextualized and accumulated through practices of ordering. Removed from the 'wild' and subject to the labour of classification and amassing, specimens become 'recombinant material' and their value thus compounded in new forms:

> By applying the knowledge, technologies and expertise concentrated within the 'centers of calculation' to the collected materials, it becomes possible to produce entirely new materials or bodies of information.
>
> (Parry 2004: 32)

Parry portrays museum collections as distillations of historical and contemporary politics of collecting. Furthermore, she analyses how museum collections are subject to transformed extensions (and concomitant deletions) of these accumulated histories as they are subject to new uses (such as the micro-sourcing of herbaria by pharmaceutical bio-prospectors in search of potentially valuable genetic resources).

Although, as it turns out, publications arising from the bird DNA barcodes did not include phylogenetic analysis,[14] what is interesting here is the curators' sense of responsibility for the epistemic uses of the material they organize and care for. They were also concerned that BOLI would 'hold onto' the sequences they had been able to rapidly process rather than make them immediately publicly available. This it was felt would prevent other taxonomists (especially students) from undertaking their own genetic research and thus stymie their career progress. The intricacies of contemporary biological materials and their simultaneous informational properties complicates the curators' sense of responsibility here. On the one hand, museum curators were concerned that BOLI scientists would let biological material move 'onwards' as data and as resource – i.e. be used in ways over which they had no control or rights. And in another way they were afraid that BOLI would hold onto data, meaning that collected biological material would be unable to be used. As suggested earlier, it is not only BOLI which is re-negotiating the precise calibrations that are implicit in the idea of a collection: in contemporary collections of natural history, implicit understanding of what can move and what must not move, into whose realm of control, and for what purposes, have to be renegotiated. Whereas BOLI seemed initially to perceive collections as gold mines of genetic material already classified and prepared and unproblematic to access, museum curators were intensely aware of their com-pounded value, as well as of their compounded specific histories and exchange relations influencing this value, thus embedding their free accessibility and use in a much denser medium of social meaning and expectation. As it happens, in this case the curators' fears were eventually allayed and by 2006, all negotiations were successfully completed, most if not all collections accessed, tissue samples mobilized and sequenced to compile a complete North American dataset of avian DNA barcodes (also housed within and accessible through BOLD).

This interlude also introduces the fact that BOLI was faced with an initial slow-down in the process of compiling the database. Whereas Parry portrays museum collections as distillations of the historical and contemporary politics of collecting, for BOLI, museums are ostensibly safe havens, liberated from the memory of centuries of labour and more contemporary politics of access. The bird collection curators we have mentioned were, however, sensitive to the fact that DNA barcoding introduces a process of 'recombining' and adding new value to genetic material through its application to new and potentially unanticipated uses. In line with Parry's argument, the curators' concerns highlight BOLI's naivety in not apprehending or complicating its need for materials ('staying with the trouble' of this, as Haraway (2010) has suggested) with a consideration of the prior labour (and associated stored value) which was invested in their initial and continued accumulation. A selective 'forgetting' appears to be in play here. The complex biopolitics of contemporary collecting were unanticipated in BOLI's assumption that it was somehow 'entitled' to seamless and unproblematic access to biological material. And museum collections turned out to be more than apolitical reposito-ries devoid of cultural and political histories. The collections and collectors alike required a sensitivity not only to histories of collection and the meanings these establish now, but also to the implicit norms and purposes of curation and collec-tion, themselves in flux given the particular new material-semiotic dimensions of natural history objects. In this context, museum collections are essentially being *re-collected* and *re-valued* by BOLI (Ellis 2009).

Our third and final example, concerning the Consortium for the Barcoding of Life's (CBOL's) involvement with the CBD, draws out reflections concerning BOLI's efforts to access fresh material in the 'wild' where it encounters more explicit and more politically intense global, national and community sensitivities.

DNA barcoding and the Convention on Biological Diversity

Since the 1992 Rio CBD, the centuries-old taxonomic practices of exploring bio-diverse areas of the world in search of unidentified organisms for research, or display, and the archiving of these in specimen collections, has become subject to more stringent regulation. The CBD's principal aims are to conserve global biodi-versity, to promote a sustainable use of its components, and to ensure the fair and equitable sharing of the benefits arising from the use of those genetic resources. Twenty years of negotiations between signatories at CBD meetings have sought in discussion, and via many formal texts, to reconcile the tension between a need to know and save biodiversity (and hence facilitate access to biological material in such areas) and a growing recognition of the values and rights of human communities who invest their own means in and live with, live from, know and have traditional collective property or ownership rights over the bio-diverse richness of their world. The CBD operates as an international legally-binding treaty providing a legal framework for the appropriate mobilization and recognition of origin of animal, plant and microbial resources. Biological resources that had previously been perceived as the 'common heritage of mankind' are repositioned (and revalued) in the CBD through legitimate claims of national sovereignty over biodiversity. In the early 1990s this was quite a radical shift with serious implications for scientific access to and use of genetic resources and has had important practical implications for BOLI's agenda (Ruiz Muller 2009).

Significantly, since 1992, a proliferation of genomic, proteomic and bioinformatics technologies and the concomitant transformation of tangible tissues into mobile and easily exchangeable units of information have transformed the scale, organization, social-economic relations, and processes of exchange and circulation which shape research and development, in ways which early CBD negotiations had not accounted for (Ten Kate and Laird 2000; Parry 2004; Oldham 2004; Ruiz Muller 2009; Vogel 2009). As Ruiz Muller and Vogel both note, implications of a new dominant scientific discourse of the genome as code (Kay 2000; Keller 2000; Thacker 2005) were scarcely recognized in the years leading up to 1992, and an ambiguity still lies at the heart of the CBD concerning the generation and global flow of information *derived* from genetic research. It should not therefore be surprising that BOLI might need to carefully navigate CBD legislation and build new far-reaching collaborations within this international policy domain to enable their own unfettered access to the most bio-diverse (and politically contentious) areas of the world. Given the scale and complexity of the CBD, the ensuing collaborations between the two 'entities' have been multiple, subtle and complex. In this section of the chapter we initiate an account of ways in which BOLI has sought to engage with the CBD, as it has sought to amplify its global sourcing of materials. We complicate this account with further detail and analysis in Chapter 5.

A systematic mobilization to globalize barcoding had been envisaged within BOLI in embryo from the very start in 2003, as described in Chapter 2. With encouragement from barcoding funders, the Gordon and Betty Moore and the Alfred P. Sloan Foundations, a de-facto consortium of research scientists and science policy diplomats had formed as the Consortium for the Barcoding of Life (CBOL) from 2003 onwards. The centrepiece of CBOL's mission was to recruit partners, service providers (including materials and access), and users, all over the globe. CBOL would provide a basis for the huge agenda of technical standardization, including laboratory and data-handling protocols, bioinformatics co-ordination, materials specifications, and so on – all of which were essential requirements for barcoding to work as a standardizing global technoscience. It was during our participation at a number of CBOL's global outreach events in Kenya, Brazil and Taiwan, between 2006 and 2008, that we (and CBOL executives) realized the gravity of constraints placed by the rapidly growing webs of national biodiversity legislation to regulate international specimen collecting, transfer and processing from such countries. Some scientists at the meetings lamented the fact that countries such as Brazil and India (biologically mega-diverse countries with notoriously stringent laws governing genetic resources) prohibited the export of genetic material beyond national boundaries, thus requiring in-country identification and genetic processing. These restrictions, which had already been affecting the progress of taxonomic research in certain areas of the world, introduced an unwelcome obstacle to BOLI's dreams of a speedy global biodiversity inventory. CBOL executives had a decision to make: either they encouraged BOLI (and later iBOL) partners and participants to comply with the increasingly intense national biodiversity CBD framed legislation – especially the parts relating to the access to resources and the sharing of benefits accrued through potential commercial development (ABS) – and simply accept the potential speed bumps; or directly enter what was for CBOL the hitherto unknown world of CBD negotiations and politics, and tackle these head on. This felt like a precarious moment for CBOL and opinions were divided.

CBOL opted for the latter course, and DNA barcoding was first introduced to the CBD community at an event organized by the Subsidiary Body on Scientific, Technical and Technological Advice (SBSTTA – CBD's prevailing scientific advisory arm) in Paris, July 2007. Before the meeting, prominent BOLI/CBOL actors believed that BOLI could and probably should circumvent the CBD, the slow and political nature of which might hinder BOLI's ambitions. This position was further supported by other taxonomic bodies, who felt that plunging headfirst into what had now been 15 years of difficult CBD negotiations would expose BOLI's relative naivety as 'new kids on the block'. BOLI would be inevitably buffeted by the complex and fiery politics around global genetic resources – an image their self-consciously scientific project would be better to avoid. Furthermore it would draw BOLI's plans to increase the global flow of genetic resources to the full attention of a community already hyper-sensitive about the politics of control, value and mobilization of these biological and virtual (informatic) materials. Conversely, other prominent CBOL/BOLI players felt that CBOL should be bolder and lead the way in promoting the role of DNA barcoding as an indispensable asset for the CBD. At this time it was thought that barcoding could be a valuable tool for rapidly clearing up disputes surrounding the origin of, and thus rights over, specimens – an issue often proving contentious amidst negotiations over disputed national claims to ownership of resources. On CBOL's decision to commit itself to working through these issues, a brochure laying out the relationship between DNA barcoding and rights of access to and ownership of genetic resources, was drafted and circulated for comment in March 2008. It was during the CBD's 9th Conference of the Parties (COP9) in Bonn in May 2008 that a direct and strategic way into the CBD for BOLI was identified that would prove distinctly advantageous in more ways than one.

It is probably fair to say that CBOL actors were unprepared for the scale of the Bonn COP9 event with its diversity of participants vigorously voicing their particular concerns, and pursuing lines of coalescence or stand-off in relationships which had been brewing for years. CBOL was inevitably a relative late-comer to the heavily contested and deeply polarized international access issue, but was quickly introduced to the fraught and distrustful relations between indigenous peoples living in biodiversity-rich areas, and nation-state governments anxious to control the new CBD regime and to exploit it for economic development. In our account below we demonstrate how BOLI (by means of the initial negotiations pioneered by CBOL) has been rather selective of the areas of politics, culture and human sensibilities generated and impinged upon through CBD negotiations (and the complex and delicate colonial histories behind them). In other words, CBOL explicitly and strategically considered what it should be sensitive towards in carving out its own route for negotiation within the CBD.

After spending some time soaking up the event, CBOL actors picked up on one particular issue of vital importance. Decade-long negotiations shaping the International Regime for ABS were it was hoped, to culminate in the signing of an international protocol in Nagoya, Japan, in 2010 at COP10.[15] CBOL's entry into this late stage of the ABS negotiations was therefore timely – that is, just in time. CBOL parties rapidly saw that the ABS framework had yet to clearly make what is a difficult differentiation between the use of genetic resources for 'purely scientific' or what it called ' non-commercial' research (seen as a public good) and their use for the more controversial category of commercial intent (in the CBD's language, 'exploitation').[16] In principle, and as desired by many parties, it was the latter which should involve more demanding and accountable mechanisms for redistribution of incurred financial benefits to provider nations or peoples, while leaving the former free of such obstacles. Without this distinction, and a corresponding loosening of access mechanisms (Material Transfer Agreements, etc.) to speed up specimen collection and transfer, what is seen as basic 'non-commercial' research (CBOL strongly believed that barcoding was in that category) would be severely hindered.

Herein lay BOLI's role, according to CBOL: DNA barcoding was ostensibly an enterprise in non-commercial science which, in the realm of biodiversity, outshone all its predecessors in terms of ambition, promise and scale. More importantly it had the potential to 'give back' to provider nations in the form of freely accessible information vital for conservation (through BOLD). Barcoding could thus be used as a flagship project with considerable persuasive power needed to clinch ongoing and difficult negotiations within the CBD concerning scientists' access to genetic resources. At the same time it could re-map the benefits that might follow from liberating 'non-commercial' scientific research from the more stringent set of controls placed upon bioprospecting with commercial intent. As a result of the CBOL's manoeuvers in Bonn (2008), CBOL actors were subsequently invited to represent non-commercial research at an ABS Working Group meeting of technical experts in Windhoek, Namibia.[17] In order to fully open out this particular debate and, importantly, to be seen to be leading the process, CBOL rapidly organized a number of discussion events, in Bonn (COP9), then later in Mexico (3rd International Barcoding Meeting) and at the Nagoya Japan CBD COP10 meeting itself, which agreed the ABS Protocol.[18] BOLI was thus moving swiftly and skillfully from naïve and late-awakening onlooker, to astute mover and shaker within the CBD.

At this point, CBOL was superseded by the International Barcoding of Life Initiative (iBOL) which was basically the same network, but with an upgrading of policy and infrastructural activities: iBOL took over the focus on these international issues of access and collection of materials.[19] Ongoing discussions through iBOL were productive and generated a list of recommendations submitted to the ABS Working Group. Among other things, these suggested a need to facilitate access to material for pure 'non-commercial' research (using distinct Material Transfer Agreements and arrangements for Prior Informed Consent) whilst simultaneously considering mechanisms giving provider countries access to any information generated on the downstream use of their genetic resources. In cases of pure 'non-commercial' research transitioning into research with commercial intent, provider countries would have the right to renegotiate the terms of benefit redistribution.[20]

Eventually a memorandum of understanding was signed in Nagoya at COP10 between the iBOL Board Chair and the CBD Executive Secretary. What was seen by BOLI actors to be a real achievement in terms of negotiating facilitated access to biological material principally for non-commercial research, could be seen to be a result of active lobbying (of course typical of the CBD context) on the part of BOLI with others like the International Council of Scientific Unions to introduce specific text into the Nagoya Protocol.[21] Significant for public relations, the 2010 Nagoya Protocol identified key common goals between the two global organizations, BOLI and the UN CBD. These common aims included: i) promoting and facilitating capacity building in species identification and discovery; ii) providing relevant processes of the CBD with biodiversity information; and iii) supporting CBD processes with respect to biodiversity targets, national biodiversity strategies and action plans, monitoring, indicators and assessments, and invasive alien species.[22] iBOL has also composed a research team of lawyers to assist barcoding scientists in the job of regulating access to material and the distribution of benefits accrued from their (potential) commercial development. The relationship has thus been framed as one of mutual benefit, a harmony publicized on iBOL and CBD websites and discussed in a widely cited article published in *Public Library of Science*, 'Barcoding life to conserve biodiversity: Beyond the taxonomic imperative' (Vernooy *et al.* 2010).

It should be clear from this description of BOLI's (and later iBOL's) relationship with the CBD that the DNA barcoders have been largely successful in navigating vital political discussions to which they arrived late in the day, and which had already been shaped by a prior politics independent of barcoding. Barcoders have acted flexibly as unfolding conditions and events required, using both political and technoscientific enterprise and skill. Through the process they have emphasized to the CBD community that taxonomy (in the form of barcoding) had not only innovated and modernized technically, with genomics and informatics as its fabric; it had also entered the twenty-first century with a degree of political flair. In doing so it was also giving taxonomy a new scientific ethos, and identity. The close working relationship crafted between iBOL and the CBD is, on the one hand, an example of an extensive and imaginative collaboration which BOLI had not previously anticipated as being so vital to its future. On the other hand, and now taking a critical distance from what Lucy Suchman would refer to as the 'artful integration' required for successful innovation, we indicate as part of our conclusion to this chapter, some less 'artful' dimensions to the efforts to access biological material, upon which neither the ABS regime nor iBOL have chosen to focus their attention. These are issues that we further develop in Chapter 5.

Reflections

All three of the cases described above are examples of ways in which BOLI's desire for genetic material has led it to build a number of globally distributed collaborations and to immerse itself, with its own agenda, in complex and open-ended constitutive natural–cultural–political entanglements. We have shown how BOLI has performed a particular imagination of its partners in the varied forms of Costa Rican parataxonomists, museum collection curators, and the politics of the CBD. We have highlighted how BOLI (and later iBOL), in its processes of active reflection and negotiation, as they have built collaborations with provider communities and countries, can be perceived as neglecting or denying the full historical and contemporary politics and human sensibilities of its collaborators.

In the first example, we highlight the fact that BOLI (via the entomologist Daniel Janzen) has made stark assumptions and performed normatively-imbued commitments about the social labour of the parataxonomists and the need to convert their social-cultural practices and their very selves. In this case, BOLI has not only denied the asymmetrical and historically deep and difficult politics inevitably thrown up through this type of collaboration; Janzen's publications articulate the view that these politics are indeed necessary for the future survival of humanity and the planet. In the case of the bird collections, we suggest that curators of the collections, scientific access to which BOLI took for granted, had to draw their attention to a certain already-established politics of use and access. In view of BOLI's visionary focus upon DNA barcoding as a 'public good' with salvational aspirations on a global scale (see especially Chapters 7 and 8), we find it deeply problematic that BOLI appears here to 'forget' the wider political implications of the deleted labour and social relations of centuries of collection, care and accumulation which are layered within collections and which are also re-used and re-valued by DNA barcoders. We find such political amnesia or inertia particularly disturbing in view of BOLI's inheritance of the politically fractured and sensitive accretion of centuries of pillage by colonial powers.

In our final example on CBOL's, and later iBOL's, head-on engagement with the CBD, we note how, thanks to its technoscientific aims and concomitant need for global materials, BOLI has had to enter a terrain already shaped by more than 20 years of sharp and shifting politics, involving a gradual reconfiguration of the global collecting landscape. Given its need for global materials, especially from the most sensitive, biodiversity-rich regions, BOLI (initially through CBOL and subsequently through iBOL) chose to pick up and run with one specific agenda presented by CBD politics – the contentious issue of the relationship between 'basic'/'non-commercial' and 'commercial' research. Entering very late in this history, BOLI lobbied ABS discussions to ensure that the ABS protocol differentiates between these modes of research in order for the CBD at least formally to remove any blockages it might have previously thrown in the path of scientific research for public good, like biodiversity protection.

In this final example, we emphasize how BOLI has played into and reproduced problematic CBD politics and power relations without attempting to destabilize or invert these. These are issues we further develop in Chapter 5. To further underscore this point here we remind ourselves of various critics of the CBD and the ABS framework in particular (e.g. Hayden 2003a and b; Oldham 2009; Vogel 2009). They argue that, through the application of ABS mechanisms, indigenous people for example, become defined as custodians and providers of biological material without considering the wider political, epistemic and ontological webs within which such material is engaged. Vogel, for example, describes the framing of current ABS arrangements as as a 'disquieting continuity of Material Transfer Agreements with the history of appropriation' (2009: 5).

In light of our more general realization explored in this chapter that iBOL has chosen not to fully consider the complex historical and cultural fabric of the context from which genetic materials are being gathered, we were even more alarmed, at the COP9 side event on ABS in Bonn, May 2009 (arranged by iBOL), to hear more than one participant question the relevance *at all* of indigenous knowledge for taxonomy and biodiversity conservation. An iBOL senior executive, confessing his confusion at the indigenous presence at the COP9, actually exclaimed, 'What is it that these indigenous people are after for God's sake!'[23] This statement was particularly troubling in view of the recent adoption of the *UN Declaration on the Rights of Indigenous Peoples*, by the UN General Assembly in September 2007, an achievement which formally strengthened the indigenous voice at the CBD.

The likely deletion from the biodiversity frame of the more complex subjectivities and histories of indigenous people including their plea for rights, dignity and self-determination[24] is particularly poignant in the context of contemporary biodiversity politics. The new Intergovernmental Science-Policy Platform on Biodiversity and Ecosystem Services (IPBES)[25] has, for example, been designed to equip biodiversity policy with a similar knowledge base as that performed for climate change policy by the IPCC. But as Hulme (2011) and Turnhout *et al.* (2012) warn, IPBES should not commit the same error as the IPCC, of reducing all framings of the issue of concern to scientific ones, thus authorizing scientific knowledge and scientific data alone. It must – in the name of democracy and hopes for sustainable futures – engage seriously and thoroughly with alternative knowledge systems and cultural framings of biodiversity in ways which far exceed the relevance of data provision.

We end this chapter by taking a step back and reconsidering the knottedness of BOLI's extended relationships as described in this chapter. As with other such ambitious technoscientific programs, these qualities are not extra to BOLI, but constitutive of it. We have argued that BOLI's distributed collaborations represent BOLI as a thoroughly de-centred assemblage. This has accentuated the potential vulnerability of BOLI as a venture posed to universalize and simplify nature operating from a clearly defined, but dependent Euro-American 'centre'. BOLI has somehow had to manage a tension between the need to harvest material generated through a diverse range of practices involving communities of complex human subjects on the one hand, and efforts to contain the hybridity and potentially disruptive centrifugal richness introduced by a proliferation of sites of innovation on the other. We suggest that the management of this process has been astute, adaptively sensitive and accommodating in some places, and disturbingly ahistorical, unreflexive and insensitive in others.

In Chapter 5 we continue to describe the ways in which BOLI was extended and consolidated as a global standard system, by finding, enrolling and aligning new partners and new materials, projects and cultures to its own aims and its own culture, while also, when necessary, flexibly accommodating some elements of difference. In both Chapters 4 and 5, we thus explore what we perceive BOLI to be, in part, promoting – an intriguing discourse of weightlessness and a universalist 'view from nowhere' (Haraway 1991, 1997; Suchman and Bishop 2000; Suchman 2002), a discourse we complicate by drawing attention to the endlessly entangled, situated and local messiness and complexity which its own expansion required.

5 Extending the barcoding frontier

We are 33 floors up and it is dark outside. We are at an evening gathering of an international scientific meeting and we feel we could be anywhere. Another tall building that we can see through the glass walls of the room is the highest building in the world, we are told. The city below seems curiously far away, out of reach. But where we are is somehow telling. This is the banqueting suite at the top of the Taiwan World Trade Centre, an economic pivot in the capital city of Taipei. We are delegates at the closing banquet of the 2nd International Barcode of Life Conference being held in this city. We chat to our neighbours around large round tables, full of delicious Taiwanese dishes. The room is beautifully decorated, the microphones for speeches are being set up, the local organizers are getting ready for the honours. Music and dancing starts, loudly, brashly – it all seems too uproarious for a scientific meeting. Taiwanese and American meeting organizers, executive scientific committee members from the Consortium for the Barcoding of Life (CBOL), and Director of the US-based Alfred P. Sloan Foundation, funders of CBOL, are called to the floor by the Taiwanese compère. They need to say a few words of address. A sense of good natured chaos characterises the small rituals that mark the end of the conference.

There's also a sense of ambiguity shadowing the evening's relaxation and celebrations – who is really hosting this large gathering? Is the lavish banquet an offering of BOLI? CBOL? CBOL's funders the Gordon and Betty Moore and Alfred P. Sloan Foundations? Or perhaps the Taiwan World Trade Centre, together with the Taiwanese state-funded scientific R&D body, Academica Sinica? How should we interpret the celebratory atmosphere surrounding the introduction of barcoding, via CBOL, to what it calls the 'South-East Asian Region'? Many cultural gestures, and translations are being made, yet the evening feels overwhelmingly un-located, somehow a-cultural, globalized. The delegates seem to make up a scientific and global 'culture of no culture' (Traweek 1988: 162).

Yet our overriding feeling is that this 'a-cultural' meeting of off-duty scientists represents a very particular kind of culture, one that is able to connect biological materials and bioinformation, or what Helmreich calls 'biology unbound' (Helmreich 2009: 280), with images of nature being protected in a pristine, overwhelmingly green, paradisiacal state. This is a culture that is confident in promoting particular forms of global scientific standardization as though this were easy, friction-free (Tsing 2005). It's a culture in which the circulation of epistemic and financial capital are intertwined in subtle ways, but where this fusion is often obscured via a predominant preoccupation with pure knowledge. It is a culture that works to align global and local subject positions through the promotion of a universal biology. It is a culture, finally, that places a high value on innovation and design but in ways that are often blind to already existing subjectivities and practices.

The above are some recollections from the closing banquet of the 2nd International Barcode of Life (BOLI) Conference, a four-day event held in September 2007 at the Academica Sinica in Taipei (see Figure 5.1). It was evidently an occasion to celebrate the hard work that had gone on over the previous four days, and to applaud some of BOLI's significant achievements. It was also a celebration, in BOLI's terms, of 'extending the frontier' of DNA barcoding to previously unreached territories. As such it was an occasion for BOLI to promote barcoding as a universal scientific practice, across this 'frontier' where divergent ways of doing taxonomy ordinarily preside. We examine this 'frontier-busting' in this chapter together with the feeling we have portrayed above: that barcoding and barcoders represent a very *particular* kind of global, a-cultural, culture that is constantly enacting connections between biomaterials, bioinformation, biostandards, nature and capital, whilst only making visible particular aspects of these enactments.

The chapter relates this portrayal of barcoding to the wider context in which this technoscience is being played out, arguing that barcoding, and its universalizing goals, are, in part, shaped by the fast-shifting terrain of the global bio-knowledge economy within which the biotechnology and genomic sciences have thrived. Vital to this economy, and visible in the global promotion of barcoding, is a fast-circulating 'roundelay' of promise, hope and on-going further visions driving scientific work, meetings, programs and projects (Olds and Thrift 2005; Thacker 2005; European Commission 2005; Fortun 2009). As for many initiatives of the global bio-knowledge economy, BOLI's world is a fast-moving, accumulative world, in which it is important to be on the brink of something new (Ong 2005: 337). We liken BOLI to Ong's description of similar 'technopreneurial' enterprises populating the global knowledge economy (Ong 2005: 344). Technopreneurial enterprises typically engage in a heady 'ecology of multiple intersecting expertises': their role is to deliberately orchestrate interactions between local and global institutions, actors and values, connecting them to new sites and labours (Ong 2005: 339).

Referring to the BOLI conferences and meetings we have witnessed, but also continuing to draw on our observations of BOLI in relation to the CBD, we want to highlight, in this chapter, an ambiguity, tension, or 'friction' also highlighted by Anna Tsing in her account of environmental activism in Indonesian forests (Tsing 2005). That is, that despite the highly interconnected, and connecting, mode in which BOLI and CBOL operate as promoters of barcoding across the globe (a mode that is entirely consistent with the wider bio-knowledge economy, biotechnology and genomics contexts), BOLI paves the way for the creation of singular, detached, biological and social *'universals'* across localities and cultures (Tsing 2005: 7; Ong 2005: 337). This promotion of universals is part of the 'frontiersmanship' of BOLI. As we described in Chapter 4, it involves a purification of the biological and social from the messiness and complexity of local contexts.

Figure 5.1 The image on the front page of the Program for the 2nd International Barcode of Life Conference in Taipei

Source: Permission to reproduce image granted by Dr Kwang-Tsao Shao of the Biodiversity Research Center, Academia Sinica, Taipei, Taiwan.

In this chapter we examine further this tendency towards purification, and the promotion and circulation of singular universals, both biological and social, as ingredients of the wider, fast-circulating, hyper-connecting world of the global bio-knowledge economy and the genomic sciences. We argue below that a disconnected and 'pure' biology, together with a homogeneous understanding of the social, is advocated by barcoders as essential to the BOLI enterprise, and forms part of BOLI's highly interconnecting 'technopreneurial' culture.

In the sections below, we look first at the global context in which barcoding is being played out – the global bio-knowledge economy and the biotechnology and genomic sciences as part of that global economy. Second, we explore BOLI's production of taxonomic/biological universals within that context. As we saw in Chapter 4, BOLI has needed to promote the idea that barcoding works unambiguously in the domain of 'basic'/'non-commercial' taxonomic research in order to facilitate smooth access to biological and genetic materials needed to fulfil the barcoding vision and fill the BOLI database. We argue that this is a particular kind of biological universalizing. We highlight the way in which BOLI underplays the complexity of genomic biology as simultaneously material and informational and we show that this is not unproblematic. Finally, we examine the way that CBOL (the institutional part of BOLI that crafts and extends the barcoding 'frontier'), works to create *not only biological but also social universals* – including global, frictionless human subjects that will adopt and promote barcoding as a standard way of doing contemporary taxonomy.

The global bio-knowledge economy and genomic taxonomy

In many ways the meeting and the closing banquet described in the opening paragraphs of this chapter were characteristic of several similar BOLI events, organized by CBOL, which we were able to attend as ethnographic observers over the period 2006–2009. These meetings were a means, as CBOL executive members[1] sometimes put it, to extend the 'barcoding frontier' in several very different parts of the world. From 2005 onwards, CBOL organized large-scale conferences such as that held at Taipei on a bi-annual basis, interspersing these with smaller 'regional' meetings that carved up the globe into manageable geographical units. By the time of the Taipei conference, such regional meetings had taken place in Durban (the 'region' being South Africa), Nairobi (East Africa) and Campinas (South America). The Taipei conference was followed by further large scale events held in Mexico City (2009) and Adelaide (2011) with many smaller events held in between.[2] All of these outreach events, organized on behalf of BOLI by CBOL, portrayed DNA barcoding as an innovation with expanding frontiers: the meetings themselves were designed to nurture the natural–technical–social collaborations that would populate this expanding geographical reach, as it extended outwards from a North American centre to encompass the globe. Each event was used to promote a sense of the urgency of barcoding as global, universal, standardized and essential taxonomic practice in the context of contemporary biodiversity loss.

We are interested here, however, in the *context* within which BOLI is pushing out what it is calling the barcoding frontier. We suggest that one aspect of this wider context is the global bio-knowledge economy, in itself a kind of new culture, born of the extension of neo-liberalism into knowledge producing domains such as museums, institutes, hospitals and universities in the 1990s. We see the global bio-knowledge economy as a fluid and politically charged domain that has spawned and nurtured the biotechnology and genomic sciences, all the while encouraging a reiterative cycling of promises, hopes and ever-on-going visions (European Commission 2005; Thacker 2005; Fortun 2009; Atkinson *et al.* 2009; Cooper 2008). But, of course, as barcoders would be quick to remind us, BOLI is not conventional genomics and so sits somewhat oddly within this context. The first thing we must do then, perhaps, is to make some distinctions.

As a new molecular technoscientific programme focused on biodiversity and environment, BOLI's innovatory program is clearly distinct from human-genomic and biomedical R&D and innovation (Atkinson *et al.* 2009). We note in passing that only a very small part of the existing literature has focused on this 'green' genomics aspect of the whole genomics-related frontier (Parry 2004; Hine 2008; Jasanoff 2005; Murphy and Levidow 2006; Levidow 2009). And from BOLI's perspective, not only are CBOL and BOLI *green* genomics enterprises, but their overt focus has been on developing a *protective* digital-DNA technoscience of biodiversity and life, rather than one which is productive or exploitation-oriented. Yet, we still maintain here that BOLI is, in part, shaped by the wider global bio-knowledge economy with its focus on finding win-win epistemic-economic partnerships. There is thus an ambiguity to BOLI's positioning within this context which we think is important as we develop below.

The global bio-knowledge economy is a fluid and unstable field in which business and scientific initiatives must lead an aggressive and speedy drive for further accumulation. As we saw in Chapter 4, this drive assumes that globalization and liberalization from regulatory interference are essential conditions for the creation of new knowledges. Such new knowledges, in turn, contribute to the bio-knowledge economy's own circulating 'roundelay' (Olds and Thrift 2005). The ever-increasing pace, self-reference, centripetal force and circularity to this dynamic is typical, and we note that, as a green, protection-oriented scientific initiative, BOLI is not at all immune to many of the pressures and urgencies faced by more commercial knowledge generating ventures.

A further quality of the global bio-knowledge economy is relevant here. We refer here to a pervasive sense of potential, an underlying anticipatory mood, that Sunder-Rajan has termed a 'future-oriented grammar' (Sunder Rajan 2006), and that Helmreich calls an 'abductive' orientation (Helmreich 2009: 172). This underlying grammar fuses the epistemic work of biologists, taxonomists and informaticians with a feeling of promise and future value. In Helmreich's terms, it transforms taxonomy into an *abductive* set of practices based upon the promise of 'unbound', 'bioinformational' forms – where abduction means the searching for entities that might, one day, have a place in a new, powerful scheme of things. Through this abductive mode, or grammar, taxonomy is newly enabled. At the same time, such a future-oriented grammar powerfully grips present imaginaries and practices. Thus taxonomy, thrown forward into the specific time-space of the global bio-knowledge economy, has already become genomics-inspired barcoding – a technoscience proper, or a set of 'distributed, heterogeneous, linked, sociotechnical circulations' as Haraway would put it (Haraway 1997: 12). These circulations have new potential: among other ways forward, they have the possibility of connecting up with an array of other scientific practices that seek out, intersect with and even co-create life forms (Chapter 8).[3]

Our observations suggest that BOLI, working within this future-oriented pulse, actively weaves its technoscience, at what it would call 'the frontiers' in to 'a net called the global' (Haraway 1997). It does so whilst declaring an unwavering commitment to biodiversity protection, and at the same time its own comportment and behaviour mimics what Haraway calls 'hypercapitalist market traffic and flexible accumulation strategies, all relying on stunning speeds and powers of manipulation of scale ...' (Haraway 1997: 13). BOLI, in other words, is part of the 'biotech' world, where life is 'enterprised up' (Strathern 2002; Haraway 1997), where 'the species becomes the brand name' (Haraway 1997: 66) and where life forms are simultaneously material and informational productions with promises attached. We suggest that it is somewhat incongruous, yet also observably the case, that BOLI, an organization explicitly connecting itself to strategies of biodiversity protection, finds itself breathing the very same air, creating and inhabiting the very same space, as the 'biotech' and genomic sciences that are more explicitly related to circuits of business and the capital exploitation of nature in sign, form and function.

What we observe is that BOLI's relationship with that hypercapitalist world, and the kind of twinning and mixing between 'nature'' and 'capital' that Franklin and Lock (2003) and Sunder Rajan (2006) have termed 'biocapital', is a troubling one. Like many forms of biotechnology, barcoding is looking to know, document and archive new life forms in ways that may or may not hold technoscientific or private commercial economic potential for the future. Like many forms of biotechnology, barcoding involves an exposure and re-arrangement of the material components (bodies, cells, DNA fragments, representations of identity, difference and function) of living things into informational components. Like other biotechnologies, therefore, barcoding exploits the capacity to materially and semiotically detach life from single bounded bodies. Like other forms of biotechnology, barcoding requires global networks of laboratories to engage precisely, and in standardized ways, with this science in order to harvest, store, freeze, amplify, sequence, classify, digitise, database and exchange life-forms in ways that preserve or prepare them for the future, be this a future of care and protection, or one of exploitation and destruction. This may be a future in which global biodiversity will be dramatically reduced, possibly to an informational rather than a material presence – a future in which species whose barcodes are banked in BOLD may become extinct, for example. It is not possible to say what the future holds for barcoded species. But what we can say is that barcoding, as part of the bioinformational world introduced in Chapter 4 (Parry 2004), is witness and enabler of life forms 'being fractionated, rendered as genealogically related, turned into networks, shored up as property, deterritorialized, and scaled down, up and over' (Helmreich 2009: 280).

BOLI and CBOL as part of BOLI, partly promotes this fractionated global biological, and partly finds itself amidst this larger scene, this 'flotsam' of life as Helmreich puts it. But at the same time it promotes barcoding as singular, universal and free from commercial connection. Barcoding needs orchestration – stable chains of translation and interlaced multiple standardizations and alignments of diverse actors, both natural and cultural, as earlier chapters have demonstrated. To fulfil its promises it also needs unrestricted access and exchange, and fast, free circulation of biological material and information through many different nodes, as Chapter 4 demonstrated. As barcoding progressively 'banks' the DNA of all life forms on the planet via the database BOLD, it comes to resemble, in more ways than one, a form of 'biocapital'. The defining language of 'barcoding' itself, as Larson (2011) has emphasized, does not help here, as it cannot escape the connotation of dealing with life and nature as commercial commodity-objects.

These observations all point to the essential ambiguities that arise when we consider barcoding as part of, and contributor to, the fast-moving, highly connective, promissory world of the global bio-knowledge economy. Barcoders do, and do not, feel part of this 'economy' which is not only an economy in the strict financial sense but also a cultural form. Barcoding, for example, disavows any potential to produce material products or profits; the CO1 gene has no obviously patentable potential in itself, the barcoding database BOLD is a publically accessible database on the World Wide Web, and the entire barcoding enterprise generally portrays itself as a purely protection-oriented mission.

Yet, and here is a similar 'friction' to those which Tsing explores, BOLI often underestimates the depth of the entangled relations in which it is implicated – including biocapital relations that have a long reach, not only into the *historical* domains of 'biology' and the life sciences, but also their *present* (via genomic molecular and informational forms), and even into larger *future* entities such as 'ecosystems' and 'nature', that will be protected via new forms of agreement and legislation predicated upon nature-capital exchanges (Brockington and Duffy 2010; Castree 2008; Robertson 2011; Sullivan 2010). As Sunder Rajan suggests, biocapital does not simply mean the 'encroachment of capital on a new domain of the life sciences'. Rather, biocapital is a form produced by the fact that the very grammar of the life sciences and of capital are being co-constituted: life has become a business plan (Sunder Rajan 2006: 283).

BOLI's universals

BOLI's business plan, however, needs to promote biological and social universals if it is to be successful in its aims to break through geographical and cultural frontiers and barcode and hence protect a global nature. Right from the very early days of barcoding, barcoders have been expert in creating biological universals. The concept, for example, that 'one gene = all species = all life', which we have introduced earlier in the book, is a brilliant example of the creative crafting of a universal biological fact. BOLI needs to form such universals in its strivings to create what we call, after Tsing, a 'frictionless global taxonomy', one in which a flow of goods, ideas, money and people can proceed unimpeded. This friction-less imaginary is one in which 'national barriers, autocratic or protective state policies' or similar obstructions do not exist, do not slow things down, and do not inhibit the freedom that comes with the mobility of 'global' people, ideas and things (Tsing 2005: 5).

BOLI needs biological universals and so it propagates them. Towards the end of Chapter 4 we documented a series of manoeuvres and negotiations undertaken by CBOL/iBOL in the context of the CBD. What we reported there was that BOLI learned quickly how to navigate the inner workings, texts and clauses of the Convention in order to create there a clearer distinction between basic/'non-commercial' and commercially-oriented science than had hitherto been included. CBOL/iBOL actors became thoroughly embroiled in the CBD in order to make this purifying move. As they did so they celebrated their efforts in different CBOL regional meetings around the world:

> The debate around the Convention on Biological Diversity has been domi-nated by bio-tech companies. They have been pressing for the CBD to lower barriers to trading and to moving species across national boundaries. What's *lost* in this debate is the value of *basic research*. In legal terms barcoding is *'beneficiary based research'*. For this kind of research there *should* be a lower barrier for sharing genetic material. That's what we [CBOL] are doing for your projects![4]

Distinguishing basic/'non-commercial' research from applied research meant that barcoding could claim innocence from the latter category. Thus, barcoders could lobby for cumbersome access restrictions on collecting and archiving to be lifted for its own 'basic research', furthering its goals as a fast, efficient, global, frictionless taxonomy. This work of purification – clearly demarcating basic from applied research – was seen within BOLI as a major achievement. What we must highlight here is that this would not have been possible without the careful extraction of a 'basic' concept of barcoding from the entangled biocapital, and nature-capital, politics of the CBD. Barcoding had to appear to be universally unambiguous, basic research oriented solely towards the identification of species. It still seems to us extraordinary that BOLI actors were successful in promoting this vision within the highly political, neoliberal, biocapital-infused context of the CBD, one of the first policy instruments to link the protection of natural and genetic resources to trade (Ten Kate and Laird 2000).

Drawing again on Tsing here, however, we note that universals, such as a purely 'basic' taxonomy, can never fulfil their promises of universality: 'they are limited by the practical necessity of mobilizing adherents' (2005: 9). This can be difficult when adherents are cognisant of history and therefore wary, as we saw in Chapter 4. Adherents, in other words, understand very well Tsing's point – that universalisms have not been politically neutral, but rather deeply implicated in questions of power and in the practices of gaining a more expansive hold on power (Tsing 2005: 9).

This sensibility to asymmetry, conveyed by Tsing, is easily seen as BOLI actors travel around the world, using ideas of a universal biology to promote barcoding. One astute commentator on BOLI's actions in this regard has levied an important critique at both the CBD and at BOLI to make clear precisely this recognition – that universals, as well as being 'progressive' can also work in deeply asymmetrical ways. The author in question, Joseph Henry Vogel, points out that the CBD, and BOLI, have not taken on board the idea, and reality, that much contemporary biological *material* is 'unbound': genomic biology is informational, codified, fractionated, as well as purely *material*, and it can be easily and widely distributed (Ruiz Muller 2009; Schei and Tvedt 2010). This overlap, as Vogel is well aware, has implications for BOLI. The material/informational hybridity which he writes about is, of course, the very currency of BOLI itself, yet at the same time the biological universals that BOLI promotes (e.g. the idea that barcoding is basic research) also deny this hybridity and its potential connections and proliferations.

What Vogel's critique does is to force BOLI not to hide behind its own biological universals, but to confront the 'unbound' potential of the biology it is working with. As many academic researchers have attested, the creation of genetic and informational 'proxies' for whole organisms has vastly complicated questions of ownership, use, re-use, and benefit. Bioinformational proxies are inherently difficult to trace over time and space, and material that has been initially collected and archived under the terms of 'basic research' can easily be shared, transported, exchanged and copied, making subsequent ownership and exploitation almost impossible to follow. As we saw in Chapter 4, the commercial interest in bio-prospecting 'in-situ' (in the countries and regions where plant and animal habitats are to be found) is currently being rapidly overtaken by companies' interest in 'micro-sourcing' – prospecting for data, or other proxies (DNA, tissue) in the existing collections already archived by museums, universities and research institutes ('ex-situ'). Parry's research shows, for example, that the corporate bioprospecting of ex-situ academic systematic collections has greatly increased since 1970 and is rapidly increasing (Parry 2004: 183; see also Ten Kate and Laird 2000). Ex-situ collections can also be used to replicate material substances (through cell culture, or combinatorial chemistry), so that biological material can be used as a means of production even if it does not have 'naturally' regenerative capacities (Parry 2004: 198; Cox 2000). Parry suggests that the *de-materialization* of collected organisms comes to weaken regulatory instruments like the CBD, instruments committed to promoting global social equity:

> [A]s the material begins to travel and unravel, with each successive decorporealization, each successive use, each remove, this commitment weakens a little further. The less material, and more informational, these resources become, the poorer, it seems, is the probability that source countries will receive any benefit or return for their use.
>
> (Parry 2004: 196–197)

Parry and others recognize that in the current genomic context, of 'unbound' biological entities, two- (or more) step processes mean that even 'basic research' (such as many forms of barcoding research), can rapidly travel, criss-crossing non-commercial/commercial boundaries (Parry 2004; Hayden 2003a and b; Plotkin 2000; Philip 2005). However, Parry also notes from her own survey of 83 academic and/ or institutional members of the Association of Systematics Collections a pervasive tendency towards denial of the depths to which biology, informatics and capital, and indeed biodiversity protection and capital, are now entwined:

> despite the increase in commercial access to and use of their collections, those working in the institutions involved seem reluctant to fully acknowledge the implications of these practices. Many display an understandable unwillingness or inability to recognize that they are now also engaged in a commercial enterprise and new trade.
>
> (Parry 2004: 182)

Politically savvy witnesses of barcoding's relationship to the CBD are alert to these dynamics, including that of denial. To come back to Joseph Henry Vogel's work on iBOL and the Access and Benefit Sharing (ABS) components of the CBD, for example, his radical proposition, graphically represented in the figures below, neatly brings together three elements:

a)   A recognition that 'genetic resource' can include not only material, but dematerialized digital information, of the kind the barcode constitutes.
b)   A recognition that biodiversity-rich, resource-poor countries often lose out through uncontrolled competition amongst one another under the ABS agreements of the CBD.
c)   A clear new role for BOLI in continuing to make freely accessible genetic information on the bio-geographical distribution of species so that entities such as Vogel's proposed biodiversity cartel (see below) can accurately and fairly negotiate the distribution of royalties from external commercial exploitation of its biodiversity resources.

Figures 5.2a and b on the following pages graphically portray this re-imagining of iBOL's role and relationships. The first image (Figure 5.2a) is the front cover of one of iBOL's promotional literatures. The second image (Figure 5.2b) has been re-worked. Here, iBOL is no longer the primary agent, 'building a world'. Rather, countries that sustain biodiverse life forms have formed a cartel, supported by iBOL's inventory of the distribution of species. iBOL now supports a 'complex' and 'entangled' rather than a 'universal' biology. Empowered by a new position of access, ownership and benefit, such countries would, in turn, support CBOL's and iBOL's efforts to create a global library of biodiversity through barcoding techniques. This controversial reversal of roles recognizes the vulnerability of CBOL, iBOL and BOLI, which, as Vogel points out, cannot be sustainably funded by grants forever.[5]

Figure 5.2a This figure was created by Joseph Henry Vogel, Professor of Economics, University of Puerto-Rico as part of a working group designed to look into the relationship between barcoding and the Access and Benefit Sharing commitments of the CBD

Source: Permission to reproduce image granted by Joseph Henry Vogel and the International Barcode of Life (iBOL) Project, Biodiversity Institute of Ontario, University of Guelph.

Figure 5.2b This figure was created by Joseph Henry Vogel, Professor of Economics, University of Puerto-Rico as part of a working group designed to look into the relationship between barcoding and the Access and Benefit Sharing commitments of the CBD

Source: Permission to reproduce image granted by Joseph Henry Vogel and the International Barcode of Life (iBOL) Project, Biodiversity Institute of Ontario, University of Guelph.

We have seen in this section that BOLI's carefully crafted biological universals radically underplay the complexity of contemporary biology and taxonomy – as material *and* informational, or as 'biology unbound'. In creating biological universals that deny this complexity, barcoding has also needed to sever a sense of attachment to the wider bio-knowledge economy and the biotechnology and genomic sciences, whilst in effect operating entirely within the cultural logics of these emergent ways of creating knowledge. The interventions of Vogel and others have served to remind barcoders to recognize the contemporary interconnectedness of biological material with biological information, to think through the multiplied and multipliable value inherent in the barcode, and to consider, therefore, how to be a technopreneurial actor that is more alert to global and north–south politics, equities and inequities. If, under the bio-knowledge economy, including the CBD, life has indeed has become a business plan, Vogel has shown that there are not just ethical choices, but issues of *justice* (Reardon 2013) that need considering as to how to conduct such business.

But the barcoding frontier, as we have suggested, involves not only the making of biological universals but also the making of barcoding subjects. We now return to the whirl of CBOL's regional meetings to look at CBOL's role in creating what we call barcoding's 'technopreneurial subjects' (Ong 2005). We analyse how such subjects are brought into being, and how, like BOLI's biological universals, they come to be contested.

Creating barcoding subjects

CBOL, as we saw in Chapter 4 through its initial negotiations with the CBD, is the face-to-face technopreneurial broker that extends the barcoding frontier across the globe. In US based business models, the actors that populate the technopreneurial enterprises of the global bio-knowledge economy are meant to imbibe, perform and reproduce a constant flow of certain management tropes – those of globalization, instrumental control, knowledge, novelty, learning, network, flexibility and urgency. The adoption of these management discourses ensure that actors turn into 'self-willed' subjects, dynamic and knowledgeable, fully capable of driving forward the entire circuitry that they are themselves creating (Olds and Thrift 2005: 275). Technopreneurial practices, Ong suggests, engender specific forms of 'technopreneurial subject', or to put it more broadly, a 'technopreneurial citizenship' (Ong 2005: 344). For CBOL's purposes this is a global citizenry of competent barcoders, who are also their competent service-providers, and users.

Whilst the typical technopreneurial subject is essentially a mobile, agile knowledge-bearing subject, detached from culture and ethnicity, and signed up to a global perspective and ontology (Ong 2005: 344), the kind of subjects that CBOL is creating are perhaps more complex. As a research initiative, CBOL recognizes very well the grandiose nature of some of its globalizing visions: 'Like the Human Genome, the goal of DNA barcoding is the construction of an enormous, online, freely available sequence database',[6] but it also recognizes its dependency on local labour and expertise: 'Like taxonomic research, barcoding is often done by researchers who are focusing on one taxonomic group in various geographic regions or a diversity of taxa in one place' (ibid.). CBOL's ultimate dream is avowedly to allow 'the products of bottom-up projects around the world to be integrated into a global initiative' (ibid.). That is, for global subjects, to sign up to CBOL-endorsed barcoding data standards in order to become *both* located *and* agile, implicated in *both* 'local' *and* 'global biology' (Keller 1995; Franklin *et al.* 2000; Franklin 2005; Lock 2000).

But how does CBOL work to align these global and local subject positions, and to reconcile local grounding with global mobility? The ethnographic observations that we describe below illustrate how compelling but also how illusory this frictionless taxonomic space is, and we come back in several instances, like Tsing, to the feeling and metaphor of 'friction' as a defining feature of CBOL collaborations across the globe. Friction, suggests Tsing, is not just a synonym for resistance: it can make global connection powerful and effective. Yet at the same time, it 'refuses the lie that global power operates as a well-oiled machine' (Tsing 2005: 6). We explore, through two stories from our field-work below, the way in which the globalizing, universalizing, quasi neo-colonial trajectory of CBOL operates in the world, and the ways that CBOL manages the tension between standardizing visions and heterogeneous taxonomic practices and subjectivities. Can innovations that have a standardizing ambition, a 'common cause' as Tsing puts it, handle the issue of difference?

Redesigning taxonomic culture

We are in Campinas, San Paolo, Brazil. The room is full of taxonomists from all over South America, greatly outnumbering CBOL executive members and BOLI affiliated scientists. The first PowerPoint presentations of the day are loaded and a speaker from the CBOL executive puts up a slide that has a reassuringly familiar list on it:

- collections, databases
- seed banks, cell-line collections
- compilations of taxonomic names
- flora and fauna surveys
- monographs, taxonomic revisions
- data repositories (e.g. of gene sequences, specimen images)
- the undigitized taxonomic literature.

The CBOL speaker immediately suggests that hardly a single item on the list presented can 'talk to' any other item on that list. 'The connections' – between voucher specimen and species name and from species name to journal publication – are, the speaker suggests, '*only* in the minds of taxonomists'. Immediately suggesting unification across existing debilitating fractures, barcoding is presented as the answer to this problem: 'Barcoding is the *first time* we have objectively and repeatedly aligned specimens and bar-codes with species names and publications'.[7]

At almost every CBOL meeting we have attended, the proceedings begin with this presentation by CBOL executive members. This, of course, has become a kind of ritual, a formula for CBOL, but it usually delivers a starkly new message to those listening, usually taxonomists who have some familiarity with barcoding techniques and technologies but are used to working with a variety of taxonomic theories, methods and apparatus. CBOL executive members do not present barcoding as just one of several possible taxonomic fronts, but as a unifying new initiative which can and will unify the fragmented disorder which prevails.

Why is observation of this opening presentation by CBOL so telling? What it shows, first of all, is that CBOL is not only advocating standardization within barcoding practices, but that it is embarking on a massive global project of standardization and interoperability. The speaker argues that the infrastructure of taxonomy is old, anachronistic, fragmented, hopelessly disconnected. And he wants to grant a special agency to barcoding for 'updating' taxonomy. He concludes: 'If you wanted to *design* taxonomy now, from scratch, you would not design it like that!'

The exchange above takes us back to the idea of the clean slate in taxonomy, even as the speaker is facing a room full of taxonomists with whom he avows to work from the bottom up. Through these initial remarks, CBOL imputes an extraordinarily high value to its own suggestions for a singular re-ordering and re-alignment of taxonomy as a global technoscientific culture. At the same time, CBOL implicitly endows an extraordinarily low value to the diverse taxonomic cultures represented in the room, epistemic cultures embodied by the taxonomists themselves. It also attributes a low value, even worthlessness, to the achievements of their more hybrid taxonomic methods, interoperable in their own historically patterned ways, and making sense 'in the minds of taxonomists'. What CBOL is suggesting, to an entirely new audience of South American taxonomists, is that their non-standardized taxonomic infrastructure and heterogeneous taxonomic subjectivities won't do. Taxonomy, CBOL is convinced, needs to be coaxed away from its own diverse, and less ambitiously 'global' cultural history and cultural practices, and to join a new purposefully designed global culture characterized by speed, global connectivity, interoperability, accumulation and growth. With an emphasis on (global) design, barcoding is put forward as the ultimate modernizing boundary object, re-ordering these disparate aspects of taxonomy, and re-situating them so that they are worked upon, not just in the minds of taxonomists, but out in the open, in publicly accessible databases where the relations and connections between different aspects of taxonomy (collections, databases, seed banks, floras and faunas, etc.) can be publically witnessed. CBOL's message, transferred through a single PowerPoint slide, quickly becomes all-encompassing: barcoding is 'a service' not only for the taxonomic community, but for biodiversity and for a new, universal community of knowers, the new global subjects of a well designed and barcoded world.

What CBOL risks, of course, in the way that it positions *itself* (and, by association, BOLI) at the centre of a future-oriented effort to redesign global taxonomy and its subjects is that this very positioning might blind them to on-going and heterogeneous developments in different parts of the world. In the coffee break that followed the PowerPoint presentation described above, a South American researcher spoke of how this blindness appeared to those in the audience:

> We have a highly developed genomics network. Our government put lots of money into genomics. I have published on barcoding but CBOL does not know this … Across Latin America we have very strong networks; we work collaboratively in networks. They do not need to know how we do this. This is how we work.[8]

In designing a new universal subjectivity for those present at the Latin American 'region', CBOL erased the possibility of understanding or working with existing taxonomic practices, and situated taxonomic subjects. BOLI's drive to design universal subjects, in effect, rendered the prior and on-going practices and identities of those subjects invisible.

Collaboration with friction at its heart?

The high value of a connected, standardized interoperable system for naming species was in many ways taken for granted by many South American taxonomists. But we attend now to some small signs of friction, occurring later on, as the standardization requirements implied by this initially attractive interoperable global vision (also seen in Chapter 3) became clear. As background to this, we should state that CBOL executives, in fact, had taken note of the development of taxonomy in the 'region' of South America. Indeed, we learnt that taxonomy was considered to be 'advanced' in this region, compared with others already visited. A sign of this was that many South American taxonomists were preoccupied *not* with the need to collect organisms and prepare specimens for analysis and archiving (voucher specimen collections seemed on the whole to be well established in museums, universities and institutes), but rather with the need to collect and store DNA. CBOL executive members, addressing participants in the meeting, and comparing them to taxonomists they had encountered on other continents, encouraged them to this effect: 'You are more focused, you are building capacity …'[9]

For CBOL however, global standardization does not work by building indigenous capacity, but rather it requires 'design', incorporating new North American standards. Heated debates, reported below, arose on one of the two days about standardizing the collection of living tissues from which to extract DNA. These debates can be seen as the resources by which CBOL could make a universal project *with* difference, or as Tsing puts it 'collaboration with friction at its heart' (Tsing 2005: 246; Graeber 2008).

Delegate A: With any specimen we must collect DNA for future studies. We should standardize that aspect of collections.

The CBOL executive member interjects:

> We don't use DMSO (a solvent that penetrates some membranes). It's genotoxic. Ethanol, also, is not good for long-term handling. The gold standard is cryo-preservation of frozen samples.

A debate ensued about this gold standard and the problem of collecting and storing frozen DNA in hot and inaccessible parts of Amazonia:

Delegate A: Alcohol-preserved tissue is just as good as frozen tissue ... So people should go ahead with collecting tissues even if they don't have freezers.

Delegate B: It's a practical issue, carrying nitrogen frozen species in Amazonia. It is impossible in Amazonia. It would take me months to get samples with that method. It's not practical at all.

The CBOL executive member disagrees, citing a six-month trip to Madagascar that used liquid nitrogen for collecting tissues. The team had food supplies dropped in every two months and with the food came a two-month supply of liquid nitrogen. He continues:

> You need to budget for that, it is realistic ... What we need to do is to establish a protocol for the collection and storage of DNA specimens ... This should be a policy that would be broadcast across the region!

In response to this came questions and protestations:

Delegate C: But what is the structure for that?
Delegate D: How can regular taxonomists be persuaded of the value of that?
Delegate E: People don't just want to be sample providers for others!

And then, a rebuff from the CBOL executive member:

> But it should just become common practice. A new standard needs to be set. It needs to be a routine, to leave a DNA sample of any specimen. This is changing culture. It should be something that is just as routine as depositing herbarium specimens.[10]

The strength of CBOL's aspiration to design standardized, interoperable epistemic and material practices can be seen in this exchange and those present are meant to imbibe this. CBOL is optimistic about the possibility of changing culture according to its own globalizing ethos. At the same time it is faced with the logics of local practices and has to consider the possibility of grafting barcoding onto existing taxonomic routines and situated subject positions. The South American taxonomists were highly sensitive about standardization and the challenges this posed to their own already routinized cultures of collecting and archiving. They defended their own situated practices as still being highly logical and valuable in their particular context. South American taxonomists also hinted at the wider politics of standardized DNA-based collecting, and questioned whose values were being promoted in this. The idea that the value in standardized DNA collections might not accrue to those carrying out the taxonomic labour was not lost on participants who were keenly alert to colonial and neo-colonial practices of collecting, the history of which we have already discussed earlier and in Chapter 4 ('People don't just want to be sample providers for others!') (Hayden 2003a; Parry 2004; Reardon 2005).[11]

Through these and other exchanges CBOL dredges up some of the oldest dilemmas of those promoting innovation, made more complex by the clashing and uneven histories of the 'hyper-developed' and developed worlds (Suchman 2002). How can innovators 'align' the need to design and 'change culture' with respect for an existing diversity of cultures? Does the *idea* of an innovation – a product, a design, a process – blind innovators to already existing situated subjectivities and practices? Is 'collaboration with friction at its heart' possible, or might new frictions give rise to incipient fractures and differences?

Conclusion

Like many genomic institutions, BOLI is propelled by a future-oriented grammar – in this case one that looks towards the protection of global biodiversity through the creation of globally standardized and frictionless biology and subjectivity. BOLI's a-cultural grounding and frame, appearing to come from 'nowhere' while claiming to see comprehensively (Haraway 1991), is, we suggest, actually a product of a very particular kind of culture, that of the global bio-knowledge economy and its technopreneurial advocates. BOLI and CBOL, its subjects (barcoding scientists) and its 'product' (globally barcoded nature), are thus very much entangled in and help to constitute contemporary fusions of nature, technopreneurialism and capital. But BOLI attempts to carve out the 'barcoding frontier', as a supposedly flawless design product of this culture, by creating a universal biology and universal subjects.

We see in this chapter that this a-cultural framing has, ultimately, to be tempered by practical enactments with 'views from somewhere' (Suchman 2002; Tsing 2005) coming into play in nitty-gritty situated negotiation of difference with new partners, accommodation of different needs, aims, conditions and constraints. Through the work of Vogel, for example, BOLI has been challenged on the way that it has leant on a universal rendition of biology that fails to recognize the complexity of the unbound entities that are, in fact, BOLI's very own currency. And similarly, in its travels through what it calls the 'regions' CBOL has met with different, varied, complex and situated taxonomic subjects who resist the global taxonomic culture proposed as an alternative to their own.

A certain blindness is perhaps propagated along with universals, as Tsing has suggested. But BOLI's creation of universal taxonomic objects and subjects (pure biology and barcoding taxonomists) are somehow more than simply naïve. They are, at least by default, deeply political, as Vogel's interventions made clear. Even amidst the entangled domain of the genomic bio-knowledge economy, these are not the only 'frontiers' it might be possible to create and we suggest there are issues of historical and contemporary justice that need to be an integral part of such a frontier. We return to this point in Chapter 9.

6 Archiving diversity

BOLD

In a chapter entitled 'Databasing the World', Geof Bowker (2005) urges us to look closely at contemporary memory practices, including the creation of biodiversity databases. The 'miracle of memory in our time', he suggests, is that 'memory practices are materially rampant, invasive, implicated in the core of our being and of our understanding of the world' (Bowker 2005: 136). Databasing the world is being done as a form of memorizing that world, archiving it, representing it and, it is often claimed, protecting it. Among myriad mundane archiving practices, we are banking seeds, databasing genomes and their DNA sequences, human culture (for example in the form of many indigenous knowledge databases), cloning and banking our pets, and creating vast material/ digital biodiversity archives. Bowker calls this a 'tumult of preservation'. This frenzy, or 'archive fever', shadowed by a quest for control or even immanence, treads a perilous line between preservation and loss (Derrida 1995; Serres 1994; Taussig 2004; Yussof 2010, 2011). Through such archives, we are also conjuring the world, suggests Bowker, 'into a form that makes it (and us) manageable' (Bowker 2005: 108). Remembering what (and who) we have excluded in this collective frenzy, may be a difficult, but essential, challenge.

What interests Bowker is how the things society commits to memory, to the database, come to be. To put this differently, how does such data come to have particular framings, exclusions or silences, contours, meanings, connections and ontologies, including forms of projection into material futures? In the knowledge and the imaginaries of different groups, what are such data thought to represent? How does infrastructure development, essential to the building of databases, influence the questions that database designers must ask and address in practice, about the kinds of objects that should be preserved, in which translations and forms, and in which relationships (Bowker *et al.* 2010; Sowa 2000)? How do such expert cultures fashion our records through the 'technical substrates' of any one epoch (statistics in the nineteenth century, generative algorithms in the twentieth)? And how do such technical substrates, combined with practices of classification and standardization, link with the archive's interior, the data themselves? How do the resulting data gain a sense of inevitability? Bowker's writing partly inspires us to enquire into the interior world of database-architecture, and the historicity and politics of their internal infrastructure. At the same time, his work can be read as a subtle warning, a sense that this same infrastructure can at best perform tautologies, or can only offer us the framing assumptions and selected facts that its designers are able to imagine. Thus, contrary to many conventional ideas about large panoptic databases (that they *reveal a world to us* in a way that might inspire, say, protection or policy), we might understand such databases as – potentially – *revealing more about the extent and limits of our own constructions and imaginations*. From this perspective, the fascinating thing about databases (or indeed large models, see Wynne 2010; Shackley and Wynne 1996; Oreskes *et al.* 1994; Keepin and Wynne 1984) is that they lay bare the extent of our cultural readiness for a world which we, despite our conventional assumptions, delimit[1] (Bowker 2000; Bowker 2005; Borges 1998).

Taking heed of these sensibilities, which invite us to explore further cultural and philosophical readings of the making of databases, we look in this chapter at the database that underpins BOLD. BOLD stands for Barcoding of Life Data Systems and is a high throughput, barcode generating, semi-automated set of interconnecting systems that is, in many ways, the pivot around which BOLI began at the University of Guelph in 2003. From 2003 onwards, Paul Hebert, scientific director of BOLI and the Canadian Barcode of Life Network, and a central figure in most '-BOL' initiatives (CBOL, iBOL and so on), built around him a skilled and industrious bioinformatics group at Guelph. This group was largely funded through grants from Environment Canada and Genome Canada[2] with whom Hebert had worked to build a close relationship. Paul Hebert has described to us in interview how Genome Canada gradually became not only funders of the Guelph Canadian Centre for DNA Barcoding of Life (CCDB), but also its advocates and partners, putting up venture capital, moving big projects forward, promoting science as a means of national knowledge economy building and as a way of 'changing the world'.[3] The group worked to consolidate the CCDB within the existing Ontario Biodiversity Institute at Guelph, overseeing the construction of new bespoke barcoding facilities, computing systems and laboratory infrastructure, and building from there a global network of ecologists, biologists and taxonomists involved in barcoding projects, including supplying the samples to meet the voracious appetite of BOLD and barcoding (the latest incarnation of which, from 2010, is iBOL, the multi-million dollar global network, introduced in Chapter 5).

The new laboratory facilities at Guelph included an initial 20,000-square-foot building built for and dedicated to barcoding, with separate floors for wet and dry analysis, together with imaging equipment, cryopreservation facilities, data systems facilities, automated robots and sequencing machines. This infrastructure, together with the humans that designed it, formed the semi-automated system that came to be known as BOLD, and through these material and human commitments BOLD soon came to represent perhaps *the* essential centre for barcoding worldwide, also acting as interface with the biodiversity and taxonomy worlds. This is what Hebert and the Guelph team wanted; and they handled the power that came on the heels of their untiring efforts with care, and subtlety: 'we should be prepared', suggested Hebert to us, 'to be donkeys for the rest of the world!'[4]

As Bowker's remarks above suggest, the commitment to sampling, databasing and making the world seem manageable, through the efforts underway at Guelph in the early 2000s, coincided with a massive global data collection drive in the bio-diversity sciences writ large (Bowker 2005: 120). Guelph's endeavours also built upon a long history of infrastructure-building in bioinformatics. Many other laboratories around the world had, by this time, succeeded in various different ways in incorporating state-of-the-art computers and softwares, and their particular powers (and particular constraints) into the collection, processing and management of biological specimens and biological data. Thus an exuberant 30-year history of bioinformatics innovation served as a backdrop to Guelph's endeavours, with several international initiatives to create large national-level databases for high-dimensional genetic data[5] having been established since the early 1980s, such as the European Molecular Biology Laboratory (EMBL), the American nucleic acid sequence database GenBank, and the DNA Databank of Japan (DDBJ) (Strasser 2008, 2011; García-Sancho 2007a and b, 2009, 2011; Cook-Degan 1994).[6]

Histories of the building of these earlier databases attest to rapid shifts in what Bowker calls the 'technical substrates' that supported them. The incorporation of the mini-computer into the laboratory in the 1970s, followed by the emergence of the micro-computer or workstation adapted as in-house computers for reduced spaces and groups of operators, in turn led to the proliferation of computerized repositories of biological information, EMBL, Genbank and DDBJ among them. (Ceruzzi 2000; Campbell-Kelly and Aspray 1996; Garcia-Sancho 2011: 71; Strasser 2008, 2011). By 2003 BOLD represented a new addition to this growing suite of vast data repositories, commonly described as: the informatization of biology (Chow-White and García-Sancho 2011; García-Sancho 2007a and b, 2009; Lenoir 1998; Hine 2006, 2008);[7] the molecularization of biology, producing the essence of biology as 'big science' parallel to physics (Kay 1993; 1995; Strasser 2011; Lenoir 1998; Kenney 1986; Gilbert 1992; Nyhart 1996); or as symbolic of a shift in biology towards high-technology 'industrialized' experimental genomics (Strasser 2008, 2011; Lenoir 1998; Kay 2000; García-Sancho 2007a, 2007b, 2011; Gauidilliere and Rheinberger 2004) – or as all these combined.

However, as in the previous chapter where CBOL, and BOLI itself, appeared both *to belong, and not to belong*, to the circuits of genomics-based biocapital and imaginaries of biocapital to come, likewise, BOLD, the global barcoding database, straddles two worlds. On one hand BOLD is an example of a new genomic-style database joining the ranks of the many other DNA sequence databases which provide the information resources and management and analysis tools underpinning modern *experimental biology*.[8] At the same time, BOLD is designed to facilitate (and discipline more extensively, through tighter if distributed coordination) the practices of collecting, describing, classifying, comparing, organizing and naming natural objects, which are so important to the study of *natural history*. In other words, whilst being an innovation of the genomic era and the information sciences, and so drawing on the technologies and ethos of those aspects of biology that are said to have overtaken the skills of natural history (Lenoir 1998), BOLD is also a database immersed in, and dependent on, the natural history tradition.

These ambiguities are complicated by two arguments. The first, which we have already rehearsed in Chapters 4 and 5, is that *natural history* practices have never been neutral with respect to commercial interest and potential. The second point is that natural history practices – a taxonomic ethos – are currently becoming more central to the practice of modern experimental and often commercial genomic sciences (Strasser 2011: 62; Kwa 2009).[9] Once again, therefore, the 'basic science' and innocent protection-oriented claims closely associated with BOLD can be read as a 'forgetting', this time of *contemporary* trends that align natural history practices more directly with those of commercially-oriented experimental biology. This can be true even if those taxonomic activities have no commercial motivations, and also are not directly functional to any such commercial intentions.

This forgetting is both a normal part of scientific culture, and also functional to its effectiveness. But it also acts to insulate the contemporary biodiversity archive from questions about its relationships with commercial drivers, and with exploitation and loss (Turnbull 2003; Taussig 2004; Yusoff 2010, 2011). Yusoff argues that the museological approach to biological life exudes a sense of care, but may actually insulate us from the reality of loss (Yusoff 2010: 88). Seen from this perspective, archiving biodiversity, is a risky, double-edged business, which may perversely inflate the likelihood and magnitude of further loss, while we think it is assisting protection. There is ambivalence here also, since it is not possible to say that barcoding, or databasing of biodiversity is in itself protecting or destroying. The causal chains and connections, thus also responsibilities, are far more complex, distant and indirect. We therefore argue below that for all the fast-paced excitement, ambition, scale and scope that we detect are part of the creation of BOLD, we need to acknowledge that this database, like all databases, in attempting to database and archive the world, enacts a very particular kind of memory and indeed a particular kind of forgetting, in making data available and accessible for its potential users.

Three tasks, and the issue of care

The designers of BOLD have summarized that this data system does three things for BOLI: it acts as a repository; it acts as a workbench; and it acts as a tool for collaboration (Ratnasingham and Hebert 2007). In this chapter we use this three-fold task-scape as a way of organizing our more cultural and philosophically oriented analysis.

First, we take up the idea that BOLD is a repository. This allows us to explore some conventional cultural ideas about scientific knowledge and databases tightly associated with this data system. We explore the idea that biodiversity databases act as neutral 'banks' for knowledge of biodiversity, that their main role is that of an accessible container or reservoir of existing knowledge to date, and that such systems, at present partially 'empty', are expectant, needing to be filled with up-to-date additions to our human knowledge of the non-human natural world. We describe how the sense of urgency to fill BOLD with DNA barcodes invests BOLI meetings and activities with a palpable tension, pace and drive to deliver, rapidly.

We explore also how such understandings of the scientific neutrality and objectivity of the database are closely connected to the idea that, ultimately, human culture will accomplish a complete knowledge of all of non-human life. For theorists trying to understand the urge to archive everything (Derrida 1995; Bataille 1991a and b; Bowker 2005; Yusoff 2010, 2011; Fava 2009), this notion of a complete record, a complete database and 'final' reference-point, is tinged with intense insecurity, and with a subtle and implicit premonition or forecast of loss, both of knowledge and of life. Thus, ironically, global seed-banks have for example been characterized by some scientists as 'seed-morgues', since they arguably create a sense of care and protection of global plant germplasm, yet cannot possibly sustain and reproduce the more dynamic human systems of culture of such seed-diversity and biodiversity in the fields, farms and markets of traditional agroecological cultures (Fava 2009).[10] In this same spirit, some concerned critics have suggested that, in their objective stasis, biodiversity databases by definition cannot contain and support life's ongoing and exuberant becoming (Turnbull 2003). In other words, control may be a false good, and one which invokes an inevitable sense of insecurity, since this loss was incurred before things even began. Thus, care (of a particular kind, perhaps) and loss are subtly joined into what is a surprising relationship within such databases. Yet the idea of biodiversity loss remains unspoken, even if it is part of what invests such enterprises with energy, and urgency. A forgetting of the possibility of loss is coupled with further imaginings: of an imagined polity in which life in all its diversity is known, is able to be known, and will therefore be protected through human agency and action. Here again we see knowledge being invested, as a matter of principle, with moral force.

   Second, we look at the idea of BOLD as a 'workbench'. Here, again, a strong ethic of *care* for the data[11] can be seen to inform the design of BOLD – both in the way that the workbench promotes 'careful' practices among its universal users, but also in the way that its internal mechanisms are mindfully chosen and maintained for their ability to produce the 'best' data outcomes. We explore in particular the idea that BOLD is a technology that facilitates 'disciplined action in time and space' (Lynch and McNally 2005). It does this through creating what Lynch and McNally call a 'chain of custody'. In BOLD designers' terms, this chain of custody is an analytical chain, sometimes also referred to as a 'pipeline' (Ratnasingham and Hebert 2007; Hajibabei *et al.* 2005; Ellis 2009). We shall go on to explore how the analytical chain or pipeline can be seen to represent a 'technology of care' over the 'career' of the samples of life that are sent to it for DNA barcoding and analysis (Lynch and McNally 2005: 297). We describe BOLD as a system designed to provide a highly specified, standardized, care-impregnated, bureaucratized, stable and accountable journey for a growing throughput of specimens as they make their way from singular specimen, to multiple and 'unbound' biological object (a fluid object that is at once cellular, organic, molecular, (bar)coded, digital, textual, photographic, cartographic (Helmreich 2009)). We show how, in combination with all these constitutive analytical chains, BOLD is a technology that produces multiples (Mackenzie 2011)[12] but that is also freighted with the moral responsibility of producing a singular objectivity (for BOLD this is a unique species identification for each specimen entered). We demonstrate how it achieves this through three different systems of tracing, witness, and accountability – 'LIMS', 'MAS' and 'IDS'. Through a focus on MAS we recognize some of the ways in which this responsibility comes to be handled.

Finally, we consider BOLD as a cyberinfrastructure for collaboration, or as Chow-White and Garcia-Sancho put it, a 'space for convergence' (2011). This collaboration and convergence does not take place in any one particular laboratory but across labs, *in silico* (Lenoir 1998) and 'at a distance', permitting the coordinated distribution of action and care for the data over space and time (Bowker *et al.* 2010; Latour 1992). Different human actors co-operate and collaborate with the BOLD data system and with each other in heterogeneous 'organized assemblages' (Lynch and McNally 2005) that span the globe – with the ultimate aim to produce clean, reproducible and accountable barcode data. BOLD facilitates the work of these humans and machines, acting as the connecting pivot, the obligatory passage point (Callon 1986: 204). Some of the most interesting collaborations turn out to be those that are concerned with the deep infrastructure of BOLD: we look here at bioinformatics ontology-building as a 'quintessential act of distribution' (Bowker *et al.* 2010), but one that works to hold together a visionary social and natural order, towards which the wider project of BOLI works. We examine one particular set of negotiations that takes place between a handful of bioinformaticians and taxonomists working together in the Data Analysis Working Group (DAWG) established by CBOL.

BOLD as repository

> BOLD is just an incredible [tool], to me, the depth, within each individual [species], of information it can accommodate, that it can go from base pair height, and phred scores from a specific base-read, a bi-directional read of a specific portion of its gene, all the way up to the specimen image, to the geographic information, to the … it's just that *overlap* of information![13]

The Barcoding of Life Data System BOLD is described by its own designers as a repository for specimen and sequence records (Ratnasingham and Hebert 2007: 2). Thus, BOLD, through its first function as a repository, enacts some of our most conventional notions of the database or archive – the idea that it is a passive kind of receptacle, a container or a 'store' (Derrida 1995: 22). And according to our interviewee, above, the BOLD repository is an amazing, exciting tool to be working with. It is not just another data-repository, but is seen to have an exceptional intensity of meaningful intelligence. As another interviewee put it, BOLD not only stores species barcodes, it stores all the information associated with that species as it moves from 'the wild' to the lab to the database – and all the way back again, at least in the minds of its practitioners. These interviewees are not alone in seeing the high-value tool created by BOLD's designers at Guelph: the particular way that the designers of the BOLD system have gone about their work has recently been recognized and highly praised by an important peer-group. Sujeevan Ratnasingham, leading from the early 2000s on the bioinformatic design of the BOLD system at Guelph, was awarded a prestigious prize in 2010 from the Global Biodiversity Information Facility (GBIF).[14]

Clearly the idea of a unitary information system that, as one of our interviewees put it, 'cleans up the mess' of long-disunited collections, representations, specimens, images and taxonomies, putting all of these 'under one registry' is highly valued by barcoders and other contemporary users of taxonomic data. In a digital scientific world, making information interoperable and indiscriminately circulable is almost an imperative of a digital scientific age, but in a domain with such urgent world-saving appeal, and with such rich historical and emotively stimulating biological resources to manipulate, it also has a powerful sense of potential. We can relate this sense of value and potential to historical notions of the 'archive'. In his book *Archive Fever*, Derrida traces the meaning of the term archive back to the Greek root *arkeion*. The arkeion is associated with the house or domicile of the superior magistrates (the 'archons') who had the power to command others. Thus, Derrida describes all archives as having this kind of 'archontic dimension'. He suggests that the idea of a singular place where all is stored (by an implicitly imagined archontic agent of control) arguably still drives the logic and semantics of our proliferating contemporary database fever (despite their silicon, virtual and placeless nature). This, he suggests, is a logic of conservation and of inscription, of putting into reserve, storing, accumulating, capitalizing, and stocking (Derrida 1995: 22; Bowker 2005; Waterton 2010b: 648; Yussoff 2010). As Derrida puts it, it is also a logic of 'consignation', an understanding of the database on which we elaborate below.

We can see then that in the traditional imagination associated with this term, the idea of BOLD as a repository conveys a sense of power to hold information in one place, organized according to a particular classificatory scheme, and from that traditionally important place, to direct affairs. But it also brings with it a sense of responsibility – to 'fill' BOLD with data to provide the archon's secure control. Pointing just outside of his small office, one of our interviewees at Guelph related:

> We've got two freezers out there, and each of them [has] 160,000 specimens of DNA extracts … it's an exciting aspect, and a tiring aspect, that … you're on top of a pile that's always growing, and it's not going to stop.[15]

BOLD is seen by this researcher, and others, as a growing data store or 'library' – which ideally requires a full complement of barcodes of all known species on the planet. This ambition is acknowledged to require considerable human labour, but it has been repeatedly explained to us that only when this library is complete will BOLD really fulfil its promised potential, as a reference point against which unknown species can be matched. In an important sense, the world itself has to be *in* the database, before it can be known (and by implication, protected). In this sense also, BOLD can be seen to play out what Derrida means when he writes of the consigning of an entire corpus, a whole system of elements. Derrida suggests:

The archontic power, which also gathers the functions of unification, of clas-sification, must be paired with what we call the power of *consignation*. By consignation we do not only mean, in the ordinary sense of the word, the act of assigning residence or entrusting so as to put into reserve (to consign, to deposit), in a place and on a substrate, but here the act of *con*signing through *gathering together signs*. It is not only the traditional *consignatio*, that is, the written proof, but what all *consignatio* begins by presupposing. *Consignation* aims to coordinate a single corpus, in a system or synchrony in which all the elements articulate the unity of an ideal configuration. In an archive, there should not be any absolute dissociation, any heterogeneity or *secret*, which could separate (*secernere*), or partition, in an absolute manner. The archontic principle of the archive is also a principle of consignation, that is, of gather-ing together.

(Derrida 1995: 3, emphasis in original)

Therefore, Derrida's idea of the archive contains a key further quality, which is that of classification and mutual ordering of the elements of the archive. Everything collected is of a single coordinated total order, with no remainder. Given BOLI's commitment to global standardization in barcoding, it should not surprise us that the concept of BOLD as a DNA barcoding library/repository of information is arguably infused with the historical, Derridean understanding of consignation. The library of BOLD, in other words, is seen as an ideal, forthcoming, complete corpus, one that is organized according to a singular gathering principle.

The library analogy performs other tasks however. First, in barcoding circles, it drives a sense of urgency amongst participating taxonomists. There is an urgency to fill the corpus, to achieve a unity of information, to gather and process speci-mens and to create a common registry of barcoded species. It is common in CBOL meetings, for example, to be given an update of the number of barcodes submit-ted to BOLD, and of the number of bona-fide validated BARCODE records held to date. Targets are often cited. The library must be filled. An ethic of common responsibility to contribute is articulated. At the time of writing it is reported on the iBOL website, for example, that in September 2010, BOLD's inventory sur-passed one million DNA barcode records for around 75,000 formally described species. iBOL predicts that, 'by 2015, this will have grown to *5 million* specimens and *500,000* species – a vast storehouse of knowledge that will be used to identify species and illuminate biodiversity' (our emphasis).[16]

Second, through the library analogy, BOLD's archive is linked to literacy, to the idea that it will 'inform', that people should be able to use and read it, and that it will emancipate global society from 'bio-illiteracy' (see Chapter 7). And so, just as this commitment to complete documentation: from specimen; to the time and place of collection; to image; to DNA profiling and naming; represents a particu-lar kind of memory practice (a total 'consignation'), it also augurs a new kind of 'applied metaphysics' (Daston 2004), an imputed ability to bind such a complete imagined repertoire of natural history objects to a future imagined global public.

In unspecified but nevertheless powerful ways, BOLD pledges a new kind of knowledge for a transformed humanity that, unlike our own, will be hungry for new ways to read and care for the diversity of the natural world.

Some researchers at the Canadian Center for DNA Barcoding of Life look forward to a time when the effort of consignation will be over, a time described as, 'once we're done'; and they assure themselves valued rewards upon reaching that time, (no longer data collection, or mapping, but *analysis*).[17] BOLD therefore imagines an 'other' future, a future of utopian wholeness in which the library is full of the DNA signatures of life, and readers are accessing this knowledge for understanding, for enlightenment, for a time of reflection upon, as well as care for, the rich diversity of the world. And so, in important ways, BOLD is a pledge, to the future, circumscribing that future, delineating and delimiting not only *knowledge of what has been* (the history of species diversity up until the present time) but also delineating and delimiting *what will happen in the world to come*. Drawing on Derrida again, we can see that the consignation of facts and artefacts is intimately linked not only to the production of the archive, to the repetition of facts, and to representation of the known past, but also to the future. In considering the nature of the archive, Derrida states:

> [T]he question of the archive is not, we repeat, a question of the past. It is not the question of a concept dealing with the past that might *already* be at our disposal or not at our disposal, an *archivable concept of the archive*. It is a question of the future, the question of the future itself, the question of a response, of a promise and of a responsibility for tomorrow.
>
> (Derrida 1995: 36, emphasis in original)

The idea of the consignation of all species on the planet within BOLD is thus complex and performative, holding together a particular imagined natural and social order, across time and into the future. But it also performs other intricacies. In her recent study of biodiversity databases, and using the work of the philosopher Bataille, Kathryn Yusoff asks whether the way that the world is ordered through archival principles acts to shape the possibilities of experience and *ethics* around biological life. Noting that the archive and archival impulses represent well-established cultural practices, and indeed prevalent attitudes towards the diversity and dynamism of life on earth, Yusoff nevertheless asks whether, by archiving diversity, we become trapped yet more tightly in an endless spiral of accumulation. Her concern is that the resulting desire – to accumulate all, to archive all – has the effect of expelling considerations of loss, care, and the destruction of non-human species. In her words: 'This is what is so paradoxical about strategies that exude *care*, but return to a ledger of *accounting* so stultified that they imprison loss in a restricted economy' (Yusoff 2010: 94, our emphasis). For Yusoff, the shift from a relation of care to one of accounting, and the very practices of accumulation driving databases of biodiversity, share the same cultural qualities as those human forces of accumulation and pretences of control which are driving biodiversity loss itself. Archives of life's diversity end up looking more like 'archives of destruction'. Yusoff quotes, 'The qualified, mechanized destruction of the Earth becomes the quantified, mechanized preservation of the Earth' (Yusoff 2010: 95, citing Stoekl 2001a: 133). Yusoff's observations echo our last insight from Derrida's *Archive Fever*. For Derrida the archive has a menacing, messianic undertone and a spectral, convoluted quality. Loss and destruction, for Derrida, inhere in the very concept of the archive: 'the archive always works,' he suggests, 'and a priori, against itself' (1995: 12).

In this section on BOLD as a repository we have examined the idea of creating a whole corpus of knowledge of global nature in a single powerful database founded not just in DNA, but in a barcode created from DNA sequences. We note in summary how this repository role structures BOLD with implicit, contradictory energies. The reassuring idea of a soon-to-be completed repository of vital data is qualified by the understanding from Derrida that the archive always works against itself. We also note further, using Yusoff's insights, that the alienation (of biodiverse global life from barcode), and global accumulation (of barcodes) driving the repository (each in itself under an ethic of care and protection) inadvertently reflect and reinforce the very same modern cultural qualities which have been destroying global life in what is now called the Anthropocene. An implicit sense of loss shadowing that of accumulation is accompanied by a denial, or amnesia, thus silence, around these contradictions.

BOLD as workbench

We turn now to the idea of BOLD as a 'workbench' (Ratnasingham and Hebert 2007). In its existence as such, BOLD begins to transgress conventional archontic dimensions, coming to seem more like a space or 'zone': where biodiversity science can be performed; where actors, human and non-human, can move about; and where translations and connections can be made from material organisms found in specific times and places to a cyber, spaceless, timeless, ostensibly a-cultural global inventory.

The workbench from the designers' perspective is envisaged as the most important element of BOLD, as well as one of the main challenges for design.[18] Characterized as 'the organization of work flows', the workbench does many things: it enables, makes traceable, manageable and accessible what Lynch and McNally, in the context of criminal forensics, call the 'career of a sample' (Lynch and McNally 2005; Lynch *et al.* 2008; Ellis 2009). During our fieldwork we repeatedly came across diagrammatic representations of this career, or journey, through BOLD. In BOLD designers' terms, it takes place within an 'analytical chain' (sometimes also called a 'pipeline'), a set of processes, steps and linkages that carry, and transform, the organism through time and space. The career of a sample is often envisaged as a journey from 'one end to the other' of this chain (Lynch and McNally 2005: 310), as Figure 6.1 illustrates.

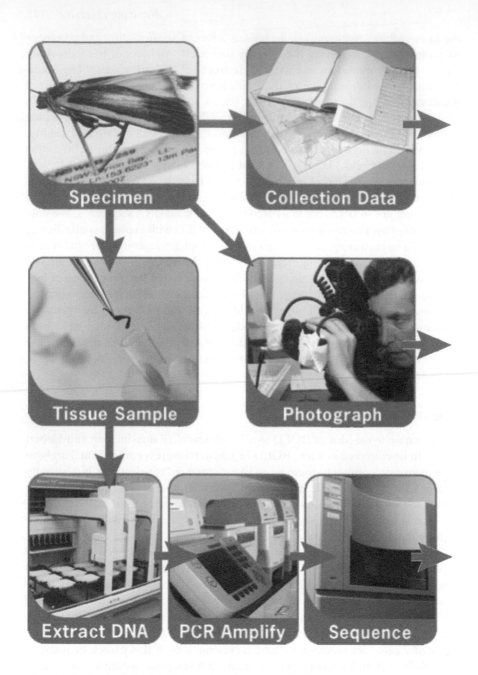

Figure 6.1 The analytical chain from sample to barcode, BOLD

Source: Created by Suz Bateson for the Ontario Biodiversity Institute, University of Guelph. Permission to reproduce image granted by Suz Bateson and Sujeevan Ratnasingham, CCDB, University of Guelph.

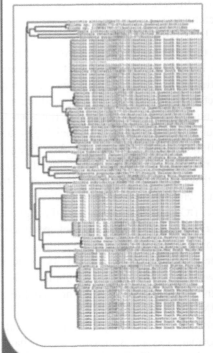

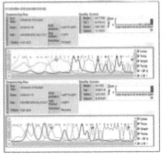

Web-Accessible Data and DNA Barcodes

Yet there is a 'misplaced concreteness' to this image which Lynch and McNally also detect in conventional understandings of the procedures underpinning criminal forensics (ibid.). This is exacerbated for BOLI by the multiple, unbound nature of the natural history specimen or 'sample' that enters BOLD. BOLD and the analytical chain within it, is a 'transductive' technology, holding within itself a complex tension, or folding, between living and non living processes (Mackenzie 2003, 2002: 193). A 'sample', for example, can enter the workflow of BOLD in at least three forms – as 'raw' specimen with field labels; as a physical array of extracted tissue in a matrix box; or as textual record or digital map of the contents of a matrix box. Samples may be entered 'at a distance' from a laboratory computer terminal in, say, Papua New Guinea, Costa Rica or other biodiverse regions of the world where barcoding projects are underway, or they may arrive as physical specimens in labelled matrix boxes to be processed by biologists and robots at Guelph. These samples enter into the workflow at different points depending on their own variable material form. Right from the start the analytical chain has to cope with multiple forms of nature–culture.

Furthermore, the analytical chain has to *take care* of all these different forms in which nature arrives. In all cases, the careers of samples have to be managed and traced: the 'chain of custody' as Lynch and McNally call it, is a technology which performs and produces a particular kind of continuity and care, required ultimately in order to establish objectivity of evidence – to be credible as such, to various unknown (but imagined) audiences/users. This calls for what Lynch and McNally call 'administrative discipline' together with 'associated systems of authority, record keeping and surveillance' (ibid.). Our fieldwork in Guelph suggested that in BOLD the entire chain is organized through a fine-tuned feeling for what each specimen needs, to ensure that each one undergoes a repeatable, traceable set of procedures in order that it may yield an objective and replicable (reliably reproducible) result. At the same time, BOLD is set up as a high-throughput data system, able to achieve a throughput of hundreds of thousands of specimen-journeys per year (when we visited Guelph in 2006, the team were aiming to achieve a throughput of half a million specimens per year). Thus, sensitivity to the organisms in their particulars is combined with an ambition and throughput of industrial proportions. In BOLD these demanding requirements are met, through three component systems of the workbench, LIMS, MAS and IDS.

The Laboratory Information Management System (LIMS) is not unusual among other high-throughput bioinformatics systems in using 'actual barcodes' – the ones you might see in a supermarket or inside library books – to determine, from the very beginning of a barcode project, what is progressively being done to a sample, creating an audit trail of information about the steps, transformations and linkages that a sample undergoes in its career through BOLD. Each of the 96 wells in a matrix box entering BOLD, for example, is stamped underneath with its own unique barcode sticker/identifier. LIMS 'pulls the informatics' out of these barcoded specimens, allowing scientists to see from a computer terminal, at a glance, what treatment any one has undergone to date. Thus LIMS traces all of the hundreds of actions that each sample undergoes, enabling individual tracking also of the multiple forms and states that all samples go through, or become, as they journey through BOLD.

The identification system (IDS), on the other hand, handles the final publication of BOLD-generated barcode sequences to NCBI and to Genbank. IDS constitutes an important part of BOLD for the barcoding researcher or project manager who wants to be given information from BOLD (identifications) as well as to contribute data to the barcoding library. We do not dwell in detail on either LIMS or IDS here. Instead we look in some depth at the Management and Analysis System (MAS), often described as the 'workhorse' of BOLD itself, which handles the many tensions involved in submitting a barcoding project through the 'central', standardizing, node of BOLD.

As Humphries *et al.* (2005) note, MAS is where all collaborative work occurs, and thus MAS treats all incoming data as part of a collaborative 'project' in which BOLD personnel, BOLD robots and any taxonomists worldwide who want to create a barcoding project are partners. MAS, in effect, handles the material, technical and cultural (intellectual property) dynamics of this partnership:

> Entries in MAS are secure, and access is granted or restricted by each project manager, so that, for example, a taxonomist may only have access to voucher and specimen records and a laboratory technician may only be able to retrieve sequence data.
>
> (Humphreys *et al.* 2005: 11)

Within MAS the workflow is characterized as proceeding in two couplets: project declaration → data submission, and data analysis → project analysis. In the first couplet, once project declaration has taken place, specimen submissions are checked both manually and by robot through a streamlined quality-control, error-detection and correction system. Checks are vital to the smooth running of BOLD and it was with some astonishment that we realized, during our fieldwork, that it is the dedicated job of two human beings working as an integral part of BOLD, to check all incoming specimen data (including mundane details such as geographical references, authors' names, addresses and so on) and to check the quality of each and every piece of photographic data entered into BOLD (digital photographs of specimens, including some that count as 'voucher specimens'). Thus, this part of MAS establishes audited disciplining flows which, in turn, create traceability and 'reversibility' of accounting.[19] The discipline of MAS is to ensure that all specimen data has to be correctly recorded before a DNA barcode is generated, and all of this data needs to be made available in case of the need to 'check back on data along the way' (Humphreys *et al.* 2005: 12).

The second set of processes (data analysis → project analysis) that takes place within MAS is where DNA-sequence alignment is performed. Here practitioners conduct myriad small, but important, actions on the sequences derived from the sample. And here, as a consequence of these many actions and translations, the tiny fragment of nature entered into BOLD becomes truly multiple (see the Technical interlude). Once nucleotide sequences have been generated for any current project, they are entered into BOLD as sequence records. Such records are then linked with the original specimen records. Only then can a new BARCODE record be generated.[20] Furthermore, once nucleotide sequences are entered into BOLD, they are automatically translated into amino acid sequences and aligned against every other sequence in the database (see Figure 6.2).

We look in some detail at this process of alignment below, but we first want to note that these activities have all had to be chosen, in some sense; or in many cases, designed from scratch as repeatable and reliable methods and processes. We shall demonstrate below that these bespoke database design choices are a way of caring for the data entered into BOLD, but also of holding together a particular, globally extensive, natural and social order which we also see as a specific kind of metaphysical imaginary. We argue that the choices made at the smallest level of detail, such as how to align sequences, matter because they are an integral part of what enables this imagined natural/social order to cohere. We draw here upon Lorraine Daston's historical research on the tacit work that the type specimen does to uphold a particular natural and social order, beyond its overt role in representing a given species (Daston 2004). Daston shows how the concept of the voucher specimen, in its present incarnation (the type specimen as a concept has changed over time) encompasses a specific kind of metaphysics – a metaphysics that holds together the transmission of knowledge, and a community of knowers, through the creation of singular designators, for things (such as species) that are innately variable, and that shift over time. The type specimen is a powerful example of an object that can support this kind of metaphysics in taxonomy, yet we suggest that it is the fact that it is at any given moment a *stable, agreed-upon, device for bringing about consent amongst a community*, rather than its originary status as a 'first specimen of a kind', that lends it this power. In a similar way we suggest that similar devices, such as algorithms buried within the very heart of BOLD, also have the power to support visions of a particular social-natural order.

Here an example is useful for clarification. We take a sequence alignment of a particular kind, as used in BOLD (see Figure 6.2). This is one based on amino acid alignment, using a particular prediction algorithm (the Hidden Markov Model (HMM)) built into BOLD. The HMM model is a prediction algorithm that predicts sequential information of various kinds (Eddy 1998). Originally developed for the statistical modelling of speech generation, its value is that it can characterize sequential information mathematically. In BOLD, the model does this for amino acid sequences at the point in the analytical chain when nucleotide sequences (ACGT …) have been derived from the specimen, but when a match of that species with other known species has not yet been attempted. At this point the nucleotide sequences have to be translated into amino acid sequences. Thereafter HMM can predict the probability that each entered amino acid sequence is the same as a previously known reference sequence. HMM produces a table of probabilities for all possible pairs (unknown with known) of amino acids and positions the model in order to calculate the highest probability path (indicating the probability that a particular sequence of amino acids is the same as another known sequence).

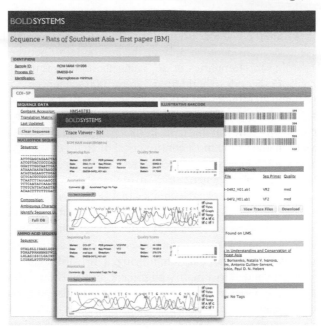

Figure 6.2 Sequence page for macroglossus minimus, including specimen details, sequencing details including links to trace files, colourized barcode representation, trace viewer and detailed stats on trace quality

Source: Created by Sujeevan Ratnasingham, CCDB, University of Guelph. Reproduced by permission of Sujeevan Ratnasingham.

These detailed steps and translations highlight several things: first, that this translation from nucleotide sequences to amino acid sequences is just one small step in a whole raft of translations that any one sample entered into BOLD undergoes; and second, that each step or translation has to be carefully considered and deliberately chosen by BOLD system designers as the best way of caring for the data. HMM, as one of many different ways of attempting alignment of sequences, was chosen by BOLD designers because translation into amino acid sequence helps in the quality control of sequence data.[21] Moreover, translation of nucleotide sequences into amino acid ones increases the number of character states from four to twenty, greatly increasing the discriminatory efficiency of the HMM, so requiring less computational power.

Such choices, bringing many contingent aspects to bear on the handling of all BOLD samples, are quite typical. Other choices in BOLD follow similar logics but all are accompanied by their own contingencies, drawbacks and tradeoffs. An example here might be the algorithm selected for the analysis of the data resulting from HMM analysis. Neighbour-joining (N-J) algorithms are used at this point in the analytical chain to calculate the distance (which equates to difference) between samples. The results of these calculations are combined to construct what is called a 'tree' (see the Technical interlude for a representation of such a 'tree'). Again, the N-J algorithm was selected from many candidate algorithms because it was considered to provide the best trade-off between speed and accuracy. According to Humphries *et al.*, using these algorithms and methods, BOLD takes only 45 seconds to build 'trees' from 1000 sequences, 15 minutes for 5,000 sequences and 20 minutes for 8,000 sequences, from any personal computer anywhere in the world (Humphreys *et al.* 2005: 14). However, each choice of algorithm, or each efficiency in this case, comes with its own tradeoffs and costs: N-J trees have been one of the most controversial of BOLD's epistemic products due to the impression which some critics believe they give, that BOLD and barcoding offers a form of phylogenetic 'tree-of-life' analysis.

Our point here is that, whereas the choices that BOLD system designers make are often made in order to simplify a highly complex, contingent and internally conflicted technical process, or to make it more efficient, the reality is that such choices are seldom without costs and sometimes even entail additional complexity. They are also formative, and not only simplifying. In the HMM case, for example, this algorithm was intended to simplify alignment. But as Humphries *et al.* note, during alignment, the underlying nucleotide sequence is mapped onto the amino acid residue alignment, ensuring that codon information is not lost. The ultimate alignment is presented as nucleotide sequences and the translation into amino acid sequences is carried out merely to enable a prediction model to work, and to speed up the alignment process. Each decision, in other words, involves trade-offs, complexities, and knock-on effects and responsibilities. In this sense, BOLD works, very impressively; but it is also fluid, fragile, contingent to some degree, and emergent. Its stability depends upon a large amount of continual, skilled and committed labour which is largely hidden to users, and suppliers. Choicemaking, tradeoffs and risks, as well as knock-on further contingencies, are an essential part of an entire system of interlinked responsibilities that BOLD designers take on, to 'curate' the data and the archive, and to take care that the analytical chain with all its diverse inputs produces what they believe to be the most reliable and robust epistemic outcome.

Finally, and here we return to metaphysics, BOLD designers make choices and take risks in order to materially uphold a particular metaphysical imaginary. This metaphysical imaginary is a particular kind of holding together, an upholding perhaps, of a particular tacit social and natural ordering, which embodies not only propositional commitments, but also normative ones. We could say it is a pledge that is enacted and made to hold through BOLD in the present, and opening out onto the future. This order's imagined future may never come about, but this is not the point regarding such a pledge, or promise (see Chapter 7). Its form is the following: as long as species identifications, wrought through transparent, standardized, agreed-upon and disciplined procedures are brought together as a complete corpus in BOLD, humanity can finally know nature; and as long as humans can know nature, in this way, they can act towards the future protection of biodiversity on the planet, thus also redeem a natural–cultural order which has fallen from the desired state of grace.

BOLD as a tool for collaboration

Collaboration, as Bowker and colleagues put it, is core to cyberinfrastructure (Bowker *et al.* 2010), which is the 'set of organizational practices, technical infrastructure and social norms that collectively provide for the smooth operation of scientific work at a distance' (ibid.: 102). This involves everyday work between scientists, informaticians, information technologists and information managers. In BOLD's case, those collaborating, as participation in CBOL meetings and visits to Guelph and The Smithsonian revealed to us, include *inter alia* taxonomists and ecologists (of many different taxa), computer scientists, information scientists, laboratory managers and bioinformaticians.

Before we turn to examine these collaborations, which, as Lenoir (1998) observes, are typically not at all fixed, but are 'constantly unfolding', we want briefly to consider a wider sense in which BOLD might be seen to be a collaborative project. As we explained in Chapters 2, 4 and 5, barcoding's extensive global network of non-human and human actors, reaches far beyond various specialist practitioners, and also includes imagined global publics as collaborative users of and contributors to BOLD. It has often been claimed by CBOL and BOLI scientists that barcoding, and BOLD as a globally accessible barcoding tool within BOLI, would 'democratize' taxonomy and biodiversity science generally, or would enable richer encounters between global publics and the practices of identifying species (Holloway 2006; Miller 2007; Ellis *et al.* 2010). In this vein, barcoding is seen as a new way of doing taxonomy that might potentially open up a much wider interface with different publics than hitherto, and transform those publics in the process. We have also heard this kind of statement many times in our fieldwork, and we have witnessed efforts of many kinds to open up barcoding 'to society' as part of BOLI. Therefore, whilst reading through our fieldnotes for this chapter, it was not very surprising to come across the following short note summarizing some of our ongoing thoughts around BOLD.

Box 6.1   Excerpt from fieldnotes, Guelph, July 2007

How BOLD innovates: NOT doing 'democracy'. NOT absorbing all wishes of communities (pluralistic, participatory) and then making the technical side more innovative.

But more: taking responsibility, daring, creative, making leaps of faith, investing and hoping people will join in afterwards – typically entrepreneurial in the information/databases world, but without the commercial implications of entrepreneurship.

BOLD in terms of 'democratization': people, demos, are following the practice; kratos, is power, leadership, a jump, a leap of faith, and an anticipation of what 'the community' will need and want. BOLD has to be ahead of the game, has to think for others, ahead of the time that people recognize what they will want.

These notes remind us of the creative, entrepreneurial innovation going on at the heart of BOLD. As we shall explore in this last section, the creative leaps of faith, and imagination of those significant others (including typical publics and specific public policy user communities) are integral to working collaboratively. But we shall also see that the responsibility for the design and inner workings of BOLD lies in very few hands. As the notes reveal, the feeling we gained at Guelph is that creativity, enterprise and innovation in BOLD, and the responsibility which this requires, are in tension with sharing, collective judgements, accommodating different visions, and with broader collaboration and inclusion. BOLD designers have to anticipate, to be 'ahead of the game', they have to 'think for others' (perhaps rather than *with* others). Hence whilst collaborations are important in many ways to BOLD, they are primarily valuable in servicing the database. A broader 'democratic' project of global collaboration, shaping not only the archiving but the imagined material protection of living biodiversity, lies in the hands of BOLD's designers; but as a future question. Having noted this distinction, we return to the inside, esoteric collaborative work, to give one example of the ongoing collaboration between taxonomists and bioinformaticians that we witnessed. This took place during one of many Database Working Group (DAWG) Meetings.[22]

We are sitting in a session at the Second Barcoding of Life Conference in Taipei (2007) that is devoted to exploring bioinformatic possibilities for barcoding. The Chair of this session opens proceedings with a pointed observation: 'Most algorithms can effectively classify 95 per cent of species. The science is in the remaining 5 per cent'. He then posed a theoretical challenge for the morning's speakers: to present to the audience, made up of taxonomists from all over the world, the different ways in which they, as bioinformaticians, had dwelt in that five per cent zone of 'science', testing and tinkering with different algorithms and data sets and comparing the results. Thus, this devoted session was deliberately organized as an interdisciplinary and collaborative encounter.

It becomes clear as soon as the speakers take their turn that the bioinformaticians, as a cultural grouping within science, have become highly skilled at rhetorically posing the kinds of question that taxonomists will recognize, as a way of communicating a common goal between them (Hine 2006). Many rhetorical questions are posed in this session, such as: 'How do we design the similarity measure?' 'How do we design efficient predictors?' and so on. It becomes clear that bioinformaticians also see it as their role to imagine and anticipate a certain set of needs. In this case, these are the needs of taxonomists working in increasingly dense bioinformatics contexts. The bioinformaticians state clearly how they are working to meet these needs within the parameters and constraints of efficacy, speed, and computing space, asking of the audience, again rhetorically: 'What do we want? Accurate and fast identification and classification, using barcodes.'

One of the morning's presentations gives a particularly vivid sense of the nature of these bioinformatic–taxonomic–barcoding collaborations taking place under the auspices of BOLI. In his presentation, one of the bioinformaticians takes great care to go into detail about the design of a similarity measure, based on the selection of salient features in specimens. Using the nucleotides ACGT, he suggests, salient features might be combinations of two letters together. Using the butterfly *Astrapes* as an example, he shows how he has used a particular algorithm to detect similarity of these salient features. He then shows how he has used the results of that calculative procedure to create a histogram. To the audience, the histogram looks like a piece of tartan fabric on the projector screen. The speaker explains that it is possible then to compute the similarity and difference between different histograms. He states clearly the benefits of this particular method: benefit 1 – no need for alignment (computationally important because alignments take up lots of space); benefit 2 – allows for the selection of substrings that are important for the identification of particular species; benefit 3 – faster than other computational methods; and benefit 4 – it is accurate.

Following his presentation the speaker explains, 'but I have not done this work with biologists yet'. Like many of the bioinformaticians, he has been trying to intuit the needs of biologists as he has worked alone on a good way to design the similarity measure. In a sense, his talk has been a leap of faith, but also an invitation to his taxonomists audience to bring some biological knowledge into his work. As he accepts questions, we are struck by the tentative but creative nature of the collaboration that CBOL and BOLI are supporting. A taxonomist in the audience, for example, asks of the bioinformatician: 'I was a bit confused about the underlying definition of a species. What is the assumption?' The bioinformatician replies: 'My definition, from a computer science perspective, is that my definition of a species is whatever you tell me a species is!' Suddenly, it feels to us, with this short exchange, the collaboration ignites, becomes exciting. Through this exchange, the collaboration opens out, in a very radical and stark way, the possibility of *choosing* assumptions, of recalibrating assumptions, of re-working the very idea of a species through this biological informatic exchange. The collaboration, seen as the interaction between this bioinformatician and taxonomist, becomes inherently liberating.

But for some the potential liberation of an interdisciplinary encounter is not the point. Another taxonomist in the audience puts forward a very different perspective:

> As gatherers of this kind of raw information … I'm speechless … We have one sequence and we say 'which of all the others does it match'? We send a whole lot of bugs in and we get bushes of branches.[23] This says to us, 'go and look at other characters!' so that, from our standpoint, your analysis is almost overkill … All we want is for the Neighbour-joining tree to be either clean or confused.

When we asked another taxonomist sitting near to us in the audience what he thought about this, he agreed with the intervention reported above, stating that it was his own common practice to 'black-box' all these issues concerning which algorithm to use: 'I just press the button!'

Reflecting on this set of collaborative encounters between one bioinformatician and several taxonomists, we can see that the interface created between these two expert cultures under the auspices of BOLI opens up a fascinating space in which it is possible to think about the orderings of nature. The construction of similarity, and hence the construction of species, for example, become a matter of open theoretical and empirical inquiry in such a new collaborative space. As discussed in Chapters 1 and 2, this is an issue very much alive in taxonomy but accentuated here by the pressures exercised by BOLI to support collaborations between taxonomy and bioinformatics, within the barcoding paradigmatic frame.

At this point we return to Bowker's idea of 'Databasing the World', or our witnessing of this making of the world being attempted *collaboratively*, as a matter of *collective care* to get, as the bioinformatician suggested, 'what we want: accurate and fast identification and classification using barcodes'. What seems clear from the interactions we have witnessed is that, in the relatively new science of identifying standard universal methodological and calculative systems for the delineation of species, a lot of different assumptions, guesses, trials-and-errors, and leaps of faith have to be made. Previous discussions about the way that scientists work together in interdisciplinary teams have often emphasized the 'boot-strapping' dimensions of this. That is, the idea that such interdisciplinary negotiations lack stable foundations, building upon fragile assumptions and connections. Even a question is an experiment in whether the other disciplinary partner recognizes it, and finds it productive. Perhaps fleetingly, the underlying contingencies are exposed. This seemed to be demonstrated in the challenges between taxonomists and the bioinformatician, in the choice of this algorithm. Yet we can also see the open negotiation of the making of another world – a world in which species delineations differ from that of a previous world. This is what Millerand and Bowker (2009) and Bowker *et al.* (2010) call ontology work. Ontology work, they suggest, is a quintessential act of distribution – taking knowledge out of a relatively closed community of practice and allowing for its reuse and reshaping by others in different fields. Whilst it could be said that ontology work is what normal disciplinary scientists do as a matter of routine, bioinformaticians use the term 'ontology' to describe their digital mathematical-algorithmic representations of specimens; and we might say that the kinds of (re)distributional ontology work occurring in interdisciplinary negotiations of the kind we describe here, are partly due to their cross-cultural character, more explicit and overt than for normal (disciplinary) science.

For ourselves as sociologists/anthropologists, it feels like a rare privilege to witness how such digital mathematical-algorithmic ontologies are made, no matter how subtly different they may be from those species definitions that are agreed upon as operative units in the present. For us, the algorithm, introduced from one community of practice into another, seems to be a powerful shaper of a potential alternative world. For, as Serres (2001) has suggested, algorithms do not act upon the data of life, but inhere within them. They inhere in the very matter of the thing, manipulating and transforming the detail of the world at its own level (Connor 2009: 23). Connor notes, 'algorithms are what Serres has called "procedural" rather than "declarative", allowing the sciences to get ever closer to the quick and quivering of things' (Connor 2009: 24; see also Serres 2001, 2009: 121). In Serres' terms: '*Et verbum caro factum est*' ('and the word was made flesh', Serres 2001: 78, cited in Connor 2009: 24; see also Kirby 2011: Chapter 4).

In this section, we have seen the way in which algorithms are, indeed, invested with a vital power in BOLD. Yet in the collaborative encounter, the weight of this, the responsibility seeming to cohere in the choice of an algorithm, seems to evaporate. The world-making which inheres in such 'technical' choices, does not seem to be sensed, and so does not have a moral presence. The diversity of perspectives in the conference room at Taipei, for example, makes such issues ungraspable, almost incommunicable (... 'I'm speechless!'). As we found out through our interviews at Guelph, BOLD's core team of designers are, in fact, those who decide how and when to subsume one algorithm with another.[24] As the principal designer of BOLD has suggested to us, it was originally thought that bar-codes would only consist of one gene, and so the first incarnation of BOLD (1.0) was 'hard-wired' to handle only one gene. In 2012 BOLD could handle multi-gene barcodes, but this innovation, required because of life's own diversity and the desire to incorporate this (Chapters 2 and 3), has required major revisions to the system. The current version, BOLD 3.0, is quite radically divergent from its predecessors but still holds the same core data elements. Even as more life, more desires for 'total' inclusion, and hence more data elements are added, a core stability to the database is maintained through the design of BOLD's internal infrastructures.[25] In this respect BOLD's designers seem to retain the authority of the 'archons', those who command. They are the ones, as our fieldnotes indi-cated, who need to be ahead of the game, and who need to practice leadership, and anticipation of what 'the community' will need and want. Theirs is the agency to effect barely perceptible subterranean shifts in the inner workings of the database (Mackenzie 2011). The world, in a sense, is theirs for the making,[26] and the col-laborators, if they are faithful to BOLD, and to the pledges built as a metaphysics into BOLD, will follow suit.

Conclusion

We began this chapter considering contemporary memory practices for the archiving of life, a focus inspired by STS scholars as well as philosophy and social theory. Through these interpretive perspectives on databases and archives, we can see that archives and databases of all kinds, including biodiversity data-bases, often reveal much about at least three interlocking elements: the limits of our own human imaginative activities, capacities and questioning; the constraints of our existing data structures; and the nature of existing commitments to manage or control nature and society. Our perspective suggests that databases are both constructed from, and tacitly re-perform, these epistemic, technical and social relations and commitments (Bowker 2000, 2005; Bowker *et al.* 2009; Waterton 2002, 2010; Waterton and Wynne 1996; Turnhout and Boonman-Berson 2011). It also suggests that database performances travel (and sometimes unravel) through what Bowker calls the 'mnemonic deep' – a layering, folding and interleaving of time that is the very fabric of memory practices. Observed through such perspectives, technologies such as BOLD begin to appear as complex, intricate systems, highly cultural assemblages of knowledge, humans and machines, designed to store and order life, and knowledge of life, in fascinating and particular ways.

Our investigations of BOLD have taken us, first, towards thinking philosophically about the work being done to create a complete standardized corpus of total biodiversity species-knowledge. We have witnessed the dedication of technoscientific practitioners to care not only for their data, but for organizing and enacting data architectures and systems. However, as BOLD grows larger, materially and in the scientific imagination, as an exciting and accessible accumulation of life and knowledge of life, the issue of ongoing biodiversity loss complicates ethics of care. Focussing attention on BOLD's relationship to accumulation and loss introduces some fundamental ambiguities and disconcertments which we think deserve further reflection. We open up some of these issues in the final chapter.

Second, we have examined the way that BOLD achieves, through its internal architectures and infrastructures (LIMS, MAS and IDS), a holding together of a particular set of ideas about the future. These are normatively weighted as well as propositional. We have described these ideas as a metaphysical imaginary that is supported through a carefully monitored series of links and translations. These links and translations connect the entering of a specimen into BOLD to a future, utopian natural and social order in which planetary biodiversity will, it is assumed, be protected. Indeed BOLD is imagined as crucial to achieving this protective human culture. It is this vision, or imaginary, in turn, that drives care and commitment to BOLD's internal disciplines and procedures, and to its community of providers and users. Stakes for the 'proper' disciplined entering of data into BOLD are high: failure to care enough to achieve this would be a failure to maintain the metaphysical imaginary at the heart of BOLD. The circularities which are in play in BOLD motivate and propel work and commitment.

Lastly, we have explored the way in which BOLD facilitates collaborations, or a kind of collective care, for biodiversity data. Such collaborations open up a window on the difficult but often exciting 'ontology work' that such a database demands. Yet we note that BOLD designers, like those of other complex memory systems (e.g. Google), are very infrequently at liberty to shift the ontological frameworks and infrastructures of the database. The many thousands of design decisions that have already been built into BOLD's infrastructure in an astonishingly short time – less than a decade – progressively act to limit the swiftness or ease with which it might shift emphasis, change direction, or alter the way that the database calculates outcomes (species boundaries, for example). BOLD, after all, is a database that aims to stabilize globally, and render available to all, what are taken to be the salient, determining features of the diversity of life on earth. As such its designers have taken on the responsibility of determining what must remain stable, and what *perhaps*, may be allowed to change, under conditions in which life itself, and our understanding of life, is constantly evolving. The BOLD designers at Guelph exercise a particular kind of power shaped by contingencies, as they work to protect the database from too much change, whilst anticipating long in advance the subtle adaptations that life itself (as it enters the database) and a growing array of users will demand, as they contribute and extract data from the database in accordance with their own relationships with life-forms. In Chapter 8 we describe our anticipation that these responsibilities and powers will shift again in the near future, as new users are encountered or recruited, and their need for forms which make the world manageable, are reimagined and folded into BOLD's emerging data structures.

7 BOLI as redemptive technoscientific innovation

In the moral and moralistic discursive productions surrounding key sectors of the life sciences, purgatorial themes and tropes retain (or, better, once again attain) a certain actuality. These concerns include a chronic sense that the future is at stake; a leitmotif among scientists, intellectuals, and sectors of the public turning on redeeming past moral errors and avoiding future ones; an awareness of an urgent need to focus on a vast zone of ambiguity and shading in judging actions and actors' conduct; and a pressing need to define a mode of relationship to these issues.

(Rabinow 1999: 17–18, by permission of the University of Chicago Press)

A chip the size of your thumbnail could carry 30 million species-specific gene sequences and brief collaterals. Push the collateral information button once and the screen offers basic natural history and images for that species, or species complex, for your point on the globe ... Such a gadget would allow access to true bioliteracy for all humanity. Such a gadget would be to biodiversity what the printing press was to literacy (and reading glasses, chairs, newspapers, the Library of Congress and the computer). The blessing of information access through such a gadget is what the taxasphere – the collective intellectual might of taxonomists, museums, collections and their centuries of literature – has within its power to offer society, global society, everyone. But will it? If it does not, wild biodiversity will continue its inexorable decline into the pit under the human heel, and the taxasphere will continue its accelerating slide into the realm of quaint esoterica shared by a very few enthusiasts who love their bugs, ferns and birds.

(Janzen 2004b: 731, by permission of The Royal Society)

BOLI scientists as prophets

Biodiversity science; the biogeographical tracing of the distribution and fate of species is, for many, a science organized around the arts and skills of prediction. · In Chapters 1, 2 and 3, for example, we describe a driving force shaping BOLI to be a sense of crisis in taxonomy fuelled by growing alarm at predicted rates of species loss combined with the slow rate of species discovery and naming. As acknowledgement of this crisis grows within biodiversity science and policy circles, so too does an anxiety that global publics fail to heed the urgency of species loss. This disregard, it is thought, results in part from failing to note the vital co-dependencies and responsibilities between humanity and the multitude of organisms with which we co-habit the planet.

We open this chapter with a curiosity for the complex meanings attributed by BOLI to curbing species loss and note that they reach far beyond a need to merely refine and speed up practices in species identification. Indeed, at the heart of BOLI's aspirations is a well developed (although selectively rather than consistently articulated) sensibility to what E.O. Wilson believes to be an innate human potential for love and kinship with(in) biological organisms through visceral and emotional connection. A sense of well-being and a harmonious future for humans and the rest of biodiversity become intimately associated and this notion of inter-species subjectivity grown through relationships of love and care, usually referred to as *biophilia*, both depends upon but also transcends a more cerebral apprehension of the differences between biological species and the human position within this (Kellert and Wilson 1993; Wilson 1984, 1991).

As such, tales of biodiversity loss and the taxonomic crisis become so much more than rates of species erasure and the documentation of this. They become imbued with a sense of lost connection and the feeling that we humans are losing our attunement to interspecies connectivity. Combined with this is regret at the loss of an inner (and, by implication, planetary) flourishing in the future. In this purgatorial model, hopes and promises for therapeutic or redemptive futures for humans and biodiversity depend upon reigniting a sense of connection with, and care for, the natural world, extra to knowledge itself. Species identification and *biophilic* redemption become mutually implicated; knowing and loving nature are inseparable. BOLI promises not only to speed up biodiversity knowledge but in the process to reunite humanity with its lost inner self once so intimately connected with nature. Most importantly for our exploration, BOLI's scientific promises thus extend from the possibilities of mundane achievement to those of transcendental and even cosmic proportions.

We realize in this chapter that we cannot attend to BOLI as ambitious innovation without carefully teasing apart its entangled pragmatic aims from those more secular-theological promises of human and planetary redemption. As the second quotation above suggests, it is clear that BOLI's hopes for contributing to curbing biodiversity loss are driven by more expansive sensual, emotional and almost spiritual hopes for human-planetary (re)connection – saving biodiversity becomes more of a cultural enterprise than initially imagined. But we are particularly interested here in what we perceive to be a tension between BOLI's transcendental hopes and promises, and the actual pragmatics of BOLI's mundane achievements. A difficult challenge facing BOLI, as we see it, becomes one of remaining alert and honest to its various aspirations some of which are clearly more closely within grasp than others! And a question relating to its challenge is whether or not BOLI finds itself accountable to its imagined addressees for its quite different types of hopes and promises. We note that, although BOLI can celebrate the banking of ever increasing numbers of barcodes in BOLD (Chapter 6), the connection between this form of achievement and the hopes for truly halting biodiversity loss and reconnecting humanity with itself and nature becomes ever more tenuous. A question we thus raise in this chapter is; what might be foregone by BOLI as its mundane and practical achievements take precedence and appear to displace the promise of salvationary futures to come, for humans and planetary life? In order to explore this question we first turn in more detail to biodiversity science's arts and skills of predicting loss whilst simultaneously positing redemptive futures-to-come.

A recent article published by Mora *et al.* (2011) emphasizes again, in familiarly alarmist terms, the yawning gap between a predicted total of the number of species on earth and the rate at which humankind is able to register and process species information in terms of presence and loss, in all global media and conditions,

> describing Earth's remaining species may take as long as 1,200 years and would require 303,000 taxonomists at an approximated cost of US$364 billion. With extinction rates now exceeding natural background rates by a factor of 100 to 1000, our results also suggest that this slow advance in the description of species will lead to species becoming extinct before we know they even existed. High rates of biodiversity loss provide an urgent incentive to increase our knowledge of Earth's remaining species.
>
> (Mora *et al.* 2011: 5)

Biodiversity science, perhaps more than any other science, enrolls us as a global public in a mourning of the devastating depletion of planetary life. Invitations to wonder at life's infinite variety of forms are tainted with a melancholic sense of what has been lost and a fear of loss to come. What is notable about the predictions in Mora *et al.*'s quotation above is that they unquestioningly assert an obligation to know (nature) before we lose the possibility of doing so. Their plea is almost to suggest that unless we act now, there will have been a copious nature belonging to the past that will have long since gone, but which we have never known. Anticipating this strange and sad future-past world, to relax about the injunction 'to know it before we lose it', in the present, would seem close to sinful. But as the complex future-past tenses of these statements suggest, an imputed moral force of knowledge seems also to combine with a premonition of its lack of worldly effect. This mood and the moral commitments that accompany it evoke the fragility and limitations of human effort together with the enormity of the task at hand – all of which is haunted by the looming specter of the turn-of-the millennium 6th Great Extinction of Biodiversity.

The urges underlying these narratives are a driving force propelling the Barcoding of Life Initiative. The opening declaration below from iBOL's website, for example (Box 7.1), gives us a sense of the value of life in the shadow of possible loss. Here, life is vibrant, sacred and useful, but our mourning for its loss is imminent. The text portrays a sense of pre-emptive regret as we realize that life is dwindling before us without being grasped by our love or knowledge.

Box 7.1 What is iBOL? What is the purpose of the International Barcode of Life?

Life is threatened

- Life is threatened with a mass extinction event rivaling any in earth history.
- Life provides critical ecosystem services, such as pollination, nutrient recycling, food and forest products.
- Life causes major economic losses linked to pests and diseases of crops, livestock and humanity.
- Life creates complex molecules, such as antibiotics and enzymes, with tremendous economic and social benefit.
- Life is largely unknown despite nearly three centuries of scientific endeavour.

Source: Project iBOL website: http://ibol.org/about-us/what-is-ibol, accessed 24 February 2012.

Hopes for bioliterate publics

Yet at the same time as forecasting rates of extinction of catastrophic proportions, and mourning species loss in advance of its actuality, we note how the biodiversity and taxonomic sciences (in this book exemplified by BOLI) offer a glimmer of hope. The possibility of a better future to come is also, we argue here, imagined in BOLI's technoscientific vision. The mantra 'we have to know in order to protect' is in part a promise that DNA barcoding, bringing with it speed, precision and accessibility, will contribute to a shared global mission to protect a planetary natural heritage which is currently being devastated. The possibility of halting or even reducing the rate of biodiversity loss is inevitably complemented by an image of return to a better ecological state and the retrieval of a lost and more bountiful natural order less marred by humanity's destructive forces. This has been articulated in a context where twenty years or more of gathering concern about biodiversity extinction has generated increasing international scientific effort, yet where there is an increasing sense of failure to influence human practices in any significant ways. Part of what is without question so attractive about BOLI is that it remains optimistic despite this depressing context.

To further illustrate this point we look to the aspirations of Paul Hebert, introduced in earlier chapters as one of the earliest, most energetic and most articulate proponents of barcoding. With characteristic vision, one that allows our imagination of barcoding's potential to fly, he described BOLI to us as 'an enterprise that promises to remake our relationship with life.'[1] When interviewed, he elaborated as follows:

Remaking our relationship with life was based on the fact that what barcoding is going to do is make it possible for anyone like me to go out into any natural environment and read the organisms that are there for the first time. That's remaking it ... I believe that there will be a significant pool of humans that will have a more intimate relationship with nature because of barcoding. I absolutely believe that and I absolutely believe it's going to happen within my lifetime.[2]

The sense, communicated by Paul Hebert's evocation, that re-making a broken collective human relationship with life will be brought about in the future through a *specific reading of nature*, sometimes referred to as *bio-literacy*, provides a perfect example of the allure of E. O. Wilson's concept of *biophilia* (mentioned above) over BOLI's imagination.[3] The concept of *bio-literacy* can be seen as a particular version and reduction of *biophilia*. Whilst *biophilia* conveys a broad affective sense of human affiliation with the natural world, the emergence of the concept of *bio-literacy* is more cognitive, promoting the idea that true love and connection is best enabled through an ability to differentiate between (and hence name and identify) natural species; to be analytically literate, that is, in their similarities, their differences, their kinship and their relations. The suggestion and hope then is that *bio-literacy*, technologically mediated by DNA barcoding, will enable a transformation from a public emotionally and ethically disconnected from nature, to one that thrives on a close epistemic (and ethical) proximity to the living natural world.

BOLI as semiotic creation

In view of the emphasis within barcoding circles upon *reading* nature and especially upon the role of the DNA barcode in facilitating bio-literacy and BOLD as both a library and dictionary for life (Chapter 6), we look carefully at BOLI's new script – the new, 'written' signs that barcoding provides. We prefer here to think of these signs as part of a particular kind of system – a 'bio-semiotic' system of a kind, since each barcode, as part of a larger system of meaning, is made up of both organism and its signs. BOLI's bio-semiotic system generates and performs a web of communication by engaging (a particular standardized DNA segment of) biological organisms, bioinformatics technologies and humans through the connective (signing) device of the DNA barcode.[4] On the one hand, as discussed in earlier chapters, the DNA barcode is designed to do a pre-determined and focused job – to signify, unambiguously, species identity and to do this in places where it matters – in situations where pest control, the regulation of protected species, or the inspiration for humans everywhere to care for nature, is at stake. For these roles, stable discriminating reference, precision and reproducibility are key qualities of the DNA barcode as sign. In these instances barcoding allows for 'samples' of hypothetical species to make a DNA based sign which (it is anticipated) will correspond to our current taxonomic understanding of those species. These signs can be seen as 'signatures' of life's diverse units – the signatories – and we refer to them as signatures from now on.[5] To express this differently, the signatories, as it were, are organisms, species. An entire host of signatories in BOLD, it is imagined, will make up 'biodiversity' as a whole.[6]

These DNA sequence-barcode signatures from nature's diverse beings can be used for various valuable purposes and as BOLI emphasizes, it is keen to develop collaborations with a growing list of users, as we discuss in Chapter 8. The DNA barcode takes on further hopeful and even redemptive meaning however. It is the hope that BOLI scientists, investing in their own scientific objects, DNA barcodes and the global infrastructure being extended to produce, organize, archive and circulate these, will extend the biosemiotic system to include publics, collectors, funders, broader user communities, future biophiliacs and a flourishing planetary biodiversity! Viewing BOLI as an expansive biosemiotic system in this way allows us to appreciate BOLI's potential for enhancing the technologically mediated communication between humans and nature. It also invites us, however, to consider the role of hope and promise in the semiotic system it has created. We turn briefly to this point below.

As semioticians have long considered, the relationship between a signature and its signatory is not usually an all-encompassing or a fully determined one.[7] To put this differently, the signature operates separately from the signatory and thus in order to understand the remit of a signature, it is important to view the relationship between the signature and the signatory as one of severance and absence (Derrida 1988: 5). Although a signature (here the DNA barcode) may signal the possible presence of the signatory (species), it will never come to fully stand in for its sensed and embodied presence. By implication, the material presence of the object being signified by the signature will always exceed the remit of the signature in ways in which the signature is not equipped to express. Another way of putting this, suggested by Derrida, is that we can really only appreciate the work performed by signatures as technologies of communication by apprehending the absence of the signatory within the signature itself. If this absence is not recognized, the reader is in a condition of reification, and the full potential of semiotic systems remains untapped. The function of the signature is then potentially cut short, and it is unable to generate a full imagination of what it is signifying – which in our case is biodiversity. The DNA barcode is only a signature of the lively and expansive presence of biodiversity itself. And as the hopes invested in the DNA barcode reach towards an emotional reconnection between humans and nature, we argue below that care needs to be taken to not allow the barcode to become a surrogate for species presence that it can only signal.

In a similar vein, the indeterminacy of signatures also requires recognition. Signatures open up pathways of interpretation and are 'performative' – the DNA barcode, for example, might generate new forms of communication and meaning only faintly anticipated by the signature's creators (see Chapter 8). Such possibilities for interpretive futures for the DNA barcode are particularly salient when we consider the DNA barcode as an expansive and hope-imbued signature. The hopeful promises associated with CO1 and barcoding as a global standardized methodology, load barcoding (as an extended biosemiotic system) with a massive responsibility. As in so many other genomic endeavors, the system becomes promissory in a way that combines biosemiotics with a sense of quasi-transcendent potential. In the following section we locate this connection of life, signature and lofty aspiration historically, describing it as typical of long-standing relationships between science, technology and religious feeling. We do this in order to think further (and somewhat critically) at the end of this chapter about the ways in which BOLI operates within a space which connects mundane achievement with redemptive promise, in subtly interesting and multivalent ways.

History of technology as redemptive potential

We began this chapter introducing the idea that new taxonomic knowledge, in the form of DNA barcoding under BOLI/iBOL, becomes an example of therapeutic and salvationary science (Helmreich 2009) for humans and nature both suffering the consequences of their disconnection. Implicit in this desired future-to-come is a reuniting of an errant humankind with (what is implied to have been) its former biosensitive self. Only then, it is surmised, will predicted rates of biodiversity loss be curbed. BOLI scientists thus task themselves with choreographing a precarious dance in which we, a global public, teeter between prophecies of apocalyptic and paradisiacal futures. Such a composition engages BOLI scientists in the task of refining their skills not only as biological scientists and as technopreneurs (Chapter 5), with claims to providing merely useful scientific knowledge and technique, but also as prophets of different possible futures, for nature and society. Their role is thus a dual one in which their responsibilities extend, albeit perhaps unintentionally, beyond a prosaic understanding of the workings of biological life, to more cosmic and socio-political processes (Toulmin 1982).

BOLI's reach towards a richer meaning and promise for earth and humanity and its infusion with both apocalyptic and salvationary visions is not of course particularly unique nor momentous for the life sciences nor for big technoscientific ventures at large (Swyngedouw 2010). Neither is the tacit relationship which we suggested above, between everyday scientific practice and prophetic visioning of future *biophilic*/dystopic worlds, unprecedented. This reach for more expansive meaning, including such quasi-divine powers of mass salvation and moral transformation, derives at least since the Enlightenment, from the widely acknowledged sense that such endeavors include, but propel us beyond, the conventional goals of mundane modes of material human betterment. Science and technology are, obliquely, effective means to take us into that other world and are thus intrinsically salvationary in spirit, and always 'reaching beyond' (Midgely 1992; Brooke 1998; Fara 2011; Noble 1999).

Scientific greats such as Newton were avowed disciples of this general creed, which gave a noble and divine role for their experimental natural philosophy. In an era where the church and science were in difficult institutional relationships, an epistemic means of transcendence was natural theology, which asserted that scientific research into nature was given meaning by its most disciplined and powerful revelation of nature's hitherto unimagined complex beauty. Its control and mastery had, since the early seventeenth-century scientific revolution, been seen as another path to human liberation on earth. Many top nineteenth-century natural philosopher scientists, whether articulate about this or more circumspect, treated natural theology as their larger scientific mission. Thus, if guided by the right moral persuasions, science was, according to this view, able to provide demonstration to the uninitiated and untrustworthy teeming masses of industrial society, of God's proper claim, through science, on their sense of debt, duty and order.

To continue in this vein, in nineteenth-century industrializing and urbanizing Britain, collective factory labour organized politically against mill-owners and their ilk to defend basic rights, and mass demonstrations were suppressed by the forces of the state. Riots and physical confrontations were a regular occurrence, and incipient social disorder was certainly a concern for elites, amongst them scientists. Infusing this uncertain human climate with a powerful intellectual discourse of a demonstrated natural order, pattern, hierarchy, and a larger beauty, created and maintained by a benign superhuman master-figure was a persuasive form of self-accounting for scientists ('natural philosophers') of that time.

At the turn of the millennium, more than a century later, new and different kinds of collective anxiety were in play, in the Global Risk Society of Beck's calling (Beck 1992, 1999). Central among these, and with emphasis from scientific authorities, were certainly anthropogenic global climate change and biodiversity loss. Also pervading these global risk discourses were expressed concerns about the widespread inability of global publics to respond, thanks to human alienation, indifference, ignorance and a moral inadequacy born of losing touch with care for nature. Thus, calls for a moral revolution to combat climate change (and its inequitable global impacts, and causes) have been as frequent as more calculatively instrumental ones over decarbonized energy or transport technologies, and so on.

In a very different domain, namely the mid-twentieth-century birth of civil nuclear power – 'Atoms for Peace' from weapons of war and mass destruction – Wynne (2010) has shown how nuclear promotional imagery and magical dreams, including the vain promise of 'electricity too cheap to meter', were being projected into society, along with representations of technoscientific nuclear elites as god-like powers, who would, as a matter of course, uphold and justify mass approval of such magical technologies and their agents. Again, as Weart (1988) has documented, in this formative post-war period, fear, hope and ambitious promise were interwoven in roughly equal proportion, in the escape from the terror induced not only by the nuclear holocausts of Hiroshima and Nagasaki, but more immediately the far greater shock induced even among nuclear experts, by early fusion H-bomb tests.

This was recent history, in the 1950s global campaign for civil nuclear power. However the persuasive power of scientific revelation and technological progress has been analysed by Schaffer (2005) in terms of more longstanding cultural need for spectacle in science's cultivation of public authority. Ezrahi (1990) has similarly described modern liberal democracy as a form of political-cultural order in which scientific reason has long gained political authority and become public legitimation currency for modern institutions beyond technoscientific ones,

through public demonstration of technological successes. Such examples emphasize not only technoscience's reach for larger meaning, but also its garnering of
credibility and license with publics expressing growing doubt and anxiety concerning the position of technoscience in the shaping of a better world-to-come.
Symbolic gestures with embodied material investments and consequences have
come to incorporate the literal, implementable, technical fix as that which will set
us on the path towards a better humanity:

> Light, knowledge, the heavens, immortality, the 'upward' movement and
> the increase of power, all sound very familiar ... But the use made of these
> symbols is quite different, indeed contrary. That use has now become wholly
> literal, offering salvation by technofix.
>
> (Noble 1999: 221)

Helmreich, writing of the teeming (and highly productive) microbial diversity
of marine life and of the harnessing of its potential by promissory pharmaceutical venture-capitalists, discusses how big life science projects, whether these be
in pursuit of human biomedical or planetary therapeutic futures, become (like
Churches) the biotechnological voices for promissory and salvationary science
feeding on growing fears of illness or devastation. 'Messages from the mud'
thus become portents, the embodied possibilities of bio-medical and ecological
therapies,

> the genes of marine microbes are texts upon which marine scientists seek
> how to inscribe commitments to reasoned thinking and teaching, responsible
> ecological stewardship and promising biotechnology.
>
> (Helmreich 2009: 30)

As various historians of science and critical theorists have demonstrated, rather
than emerging as a result of merely coincidental developments, salvationary
aspirations of technoscience share a propelling force of a 'chronic sense that
the future is at stake' (Rabinow 1999: 18), and they have been contingent upon
and entwined within changing theological doctrine within Western Christianity
(Szerszynski 2005). Particularly relevant here for appreciating some of the more
subtle dimensions of the exalted role of technoscience are the ambiguities within
Christian theology concerning both the means towards and the *location* of sacred
or paradisiacal futures-to-come. As Szerszynski (2005) and others trace, it is not
only a question of whether the sacred is of this or the next world; it is also a
matter of whether humanity can be liberated by virtue *either* of contemplation
and detachment from this worldly sensory engagement, *or* of fully embodied and
intentional material action.

   Noble (1999) has suggested that imagined futures as impending dystopias but
with paradisiacal potential have fueled and justified the 'practical and useful arts'
of technological design since at least the rise of millenarian Christianity in twelfth-
century medieval Europe. Millenarianism introduced to Western Christianity the

belief that a sinful humanity and its lost divinity could be reunited not by contemplation of the divine beyond this world, but by actions of human intervention on earth. A promised future-to-come and the joint ideas of an improved social and natural order would achieve earthly paradise. As Szerszynski further notes, this theological shift regained traction in post-Reformation Protestantism and he argues, in line with Noble in a way pertinent to this chapter, that it is this potential for redemption on Earth that opens the way for technological mediation; technology can bring about a sensually aware, bodily present divine re-connection on earth rooted in this worldly intention and (technoscientific) action in the here-and-now:

> technology as a phenomenon is further transformed in the modern period, as the transcendent axis, radicalized by the Reformation and modern science, is itself intro-jected into the empirical world. During this period sovereignty comes to be conceived in terms not of transcendence and difference, but of the maintenance of society's own immanent coherence.
>
> (Szerzynski 2005: 58)

Thus, one might say that as the divine and transcendent being with reach beyond the human is secularized through modernization processes, the material and epistemic capacity of technoscience to demonstrate systematic 'reach beyond the human', may have been reciprocally invested with quasi-divine powers, which would tend to imbue technoscientific culture with just the manifest qualities which analysts such as Noble, and Szerszynski, explain. In a related sense, a new biological understanding of the sacred qualities of life on earth emerges and technology thus becomes an active counterpart in organizing life itself and in so doing takes on an enabling role in bringing the transcendent divine ever closer to this worldly existence.

Importantly for our analysis, the nostalgic sense that humanity may be reconnected with a temporarily misplaced divine self – traced back to Millenarianism – resonates with BOLI's this-worldly *biophilic* redemptive imaginary. But perhaps it is also worth remembering that debates within occidental Christianity concerning the ambiguous nature of empirical or transcendental paradise, and between contemplative or materially active routes to liberation, also present a tension unresolved within BOLI's aspirations and promises. A *bio-literate* or *biophilic* humanity is one which lovingly reconnects with nature precisely through a contemplative reading of species similarity and difference – a contemplation not only of multitudinous diverse beings but of their identity and bio-geographical distribution. It is this combination of care and contemplation which is expected to provide a therapeutic future for biodiversity itself. The future of biodiversity indeed depends upon a welding of contemplation, care, knowledge, modesty and action (all of which are shadowed by a belated sense of remorse, and hopeful potential for transformation). We return to this subtle entanglement between distance and engagement below.

BOLI as expansive cosmology

To summarize from this brief historical foray into the quasi-religious role of technology from the twelfth century to the present, although God may (or may not?) have been dismissed over time in science and society as the ultimate reference-being, we have illustrated a general urge to invest larger social and spiritual meanings in the ordinary pragmatic and finite routines of scientific activity (in the form of improving social and moral order, awe-inspiring spectacle, and future therapies). Technoscience in all of these examples has taken on the mantle of at least partially delivering mankind from suffering and fear with dreams of fulfillment and redemption. This narrative has some profound similarities with the DNA barcode's genomically-referenced redemption which BOLI has articulated as its promise. BOLI's *biophilic* hopes are morally charged, to combat the threat brought against biodiversity (and human fulfillment), by indifference and moral inadequacy. iBOL's web-based mournings in Box 7.1, and the fervent statements of BOLI scientists, are only examples of a more generally growing anxiety about the precious and finite existence of human and other organisms on the planet, and about the failure of their own trade, taxonomic science, to have made any difference to this impending demise (see Chapter 1).

BOLI's crusade for taxonomy, all species, human benefit and the planet is a laudable one. This may or many not be envisaged as a spiritual venture and we are not suggesting that BOLI is explicitly theological. But we do argue that BOLI is intuitively aware of humanity's imaginative needs for mythological meaning, and there is something powerfully performative about the imaginary of hope and promise we describe above. Indeed, we suggest that the extremely ambitious scope of global public enrolment that BOLI embodies requires almost by definition such mythological transcendent reach. Human and cosmological fragility has long been a vital resource for scientific research and intervention (Midgely 1992) and BOLI, by our account is a quintessential example of a public salvationary science.

Divine promises of technoscience in general, and of BOLI specifically, tantalize the imagination and bring forth a sense of anticipation as to what human–natural–divine re-connection as part of an earthly attainment of liberation might engender. By offering technoscience as an incorporation of, and conduit to, the sacred (or perhaps to its secular equivalent, supra-rational transformation), the pathway seems an open and generative one. It is also one which by definition is suggestive of indeterminate ends and possibilities. It is worth reiterating here a point we have raised during this chapter concerning the (historical) need for spectacle and public witnessing of technoscience's achievements. On the one hand BOLI's promises of *biophilic* redemption and related restoration of a flourishing biodiversity are presented in spectacular tones as scientists voice a commitment to a more expansive meaning for taxonomy, via the DNA barcode. On the other hand and as we explore below, BOLI's record of actual achievement to date, and indeed its plans for future activity, is much more prosaic in nature. Indeed, explicit routes of accountability for BOLI in the eyes of science and its publics remain firmly within the mundane and pragmatic successes denoted by precise and faithful species differentiation and the accessible banking of ever more (accurate) DNA barcodes in BOLD.

It is thus interesting here to note the concern voiced by historians of science and critical theorists already cited above, that science and technology might – somewhat paradoxically – oscillate between two mutually exclusive contraries: taking up boundless, expansive and far-reaching salvationary aims and related cosmological commitments; but also having to be preoccupied as their public trade with the more routine dimensions of science as a much more bounded and functional system of instrumental knowledge- and technology-development. The question of the relations of these apparently mutually contradictory cultures, both as elements of technoscience as we know it, is one which interests us here.

Toulmin, for example, has argued that – more as a function of its mode of organization and practice than of its substantive insights – modern science (from the seventeenth century onwards) brought about a demise of cosmological modes of thought. With modern science came a loss of contemplation about connection, continuities, and relations between separate elements of the cosmos, from the molecular to the interplanetary. Like Ellul (1964 cited in Szerszynski 2005), he notes that to science, the field of interest itself tends to be bounded; so that not only did *disciplines* become modern *modus vivendi*, but dispassionate mechanistic modes of *imagination* were established over time, and tightly bounded fields with bounded imaginations, bounded questions and bounded means of observation and measurement came to define knowledge making. Cosmological integration was thus no longer imaginable and moreover, the new scientific ambitions and norms of ever-increasing precision and streamlined functioning were themselves canonized as the realm of the sacred.

Szerszynski, drawing on Ellul describes the changing denotations of the technological and compares traditional 'technai' with modern technique where technai are inclusive practices towards the ends of 'human flourishing, incorporating ideas of beauty, justice and contemplation' (Szerszynski 2005: 59) and modern technique becomes more 'self-directing, a closed self-determining phenomenon' (ibid.: 59), 'Fundamentally, technique becomes an end in itself, in which elements are functional – adapted not to specific ends but to the needs of the system as a whole' (ibid.: 59).

This suggests that we might expect to see a contemporary technoscientific landscape where reductionist mechanism and self-justifying epistemic principles of precision and control dominate so powerfully as ends-in-themselves that any larger meta-scientific imaginaries of the kind we identify with BOLI are deleted, and would only exist if at all, as fragmentary and marginal historical relics of a by-gone age of more cosmological ambitions. However, this is not at all what we find with BOLI, and neither is this an abnormal technoscientific culture in the age of genomics and big biology. Yet the expressed possibility that the secular-theological promises of technology for individuals and society at large somehow become reduced and sucked into a self-referential promise of perfection of technology itself divorced from its broader potential social partners and its broader aims of social and ecological betterment are a risk. They can also be seen not as alternative states of being, but as essential and paradoxically contradictory components of a complex and ever-emergent global technoscience, BOLI, whose future character will be shaped by the interplay of these contradictory elements. In previous chapters we have documented BOLI's efforts to perfect a streamlined DNA barcoding system accountable and useful to scientific and public communities alike. A question we raise in this chapter is to what extent might BOLI run the risk of dazzling itself with its own potential successes in universal precision and standardization? And in so doing might it forget the larger, more sensual and ambiguous connections promoted by the very notion of *biophilia*, the idea of paradise on Earth, and the nurturing of 'technai', the very connections that might lead us towards the promised human–non-human flourishing?

Operating between earthly ecologies and intangible futures

BOLI's investment in generating a semiotic system in which the DNA barcode as biodiversity's signature is weighted with complex meaning presents us (and BOLI) with some potentially fruitful tensions between mundane practical achievement and extravagant promise for open-ended transformative futures-to-come. We remind ourselves here of previous chapters where we empirically document the prosaic activities of specimen sourcing, the purification and digitization of DNA sequences, the enrolling of a global community of taxonomists, laboratory managers, technicians and bioinformaticians. We emphasize in particular (especially in Chapters 2 and 3) the ways in which BOLI's focus is aimed towards improving the stability, precision and universality of the barcode in order to equip it with simple but universal taxonomic potential, at least for identification. And as we shall see in more depth in the next chapter, much of the potential of BOLI lies in a range of very practical applications which continue to expand and flourish amongst the various 'user' communities of fast taxonomic services. A reminder for now of these pragmatic dimensions of DNA barcoding comes from a glimpse at iBOL's quarterly 'Barcode Bulletin' published on its website. iBOL's achievements are regularly updated and include the following: 'FDA using barcoding to spot fish fraud (nine labs now testing fillets for mislabelling – March 2012); What's in your tea bag? (three high school students use barcodes to find the answer – October 2011); Tracking the tsetse fly with DNA (barcoders are hot on the trail of the notorious disease vector – July 2011)'. These examples, albeit containing some important practical uses, which we know are constantly being extended, seem a far cry from the extravagant and somewhat abstract promise of human and planetary salvation through BOLI. Most important for our concerns in this chapter however is the reminder that BOLI scientists bring together boundless expectations in terms of what the DNA barcode can bring to the taxonomic and biodiversity sciences and to human society at large, with what are, on the other hand, extremely bounded outcomes and realities. In a practical sense, barcoding barely accelerates our knowledge of (unknown) biodiversity nor does it alleviate destruction. It simply provides a tool for re-archiving information of existing known species and for rapid identification of the same at sites in which identification is in doubt – in other words, addressing the needs and expectations of the USA FDA makes the prospect of the promised planetary salvation seem very distant indeed.

The following extract from an interview with a researcher from the Canadian Center for DNA Barcoding, in a rather mild sense, and tempering some of the more profligate claims of his colleagues, expresses well a pairing between informational infrastructure and an aspiration for broader transformative salvationary potential:

> So if you're concerned … about the disappearance of biodiversity in our time, the management of ecosystems and resources upon which humans depend, then you cannot help but come to the conclusion that we need to harness technology in ways that are beneficial for our environment and the biodiversity on earth. And barcoding has a role to play in there just as surely as bioinformatics and geo-spatial positioning systems. All of these things are coming together in a project of information management of unprecedented scope. *So I don't say that barcoding in itself is going to save the world but it is a very integral component to managing information.*[8]

We are not suggesting here that BOLI scientists are unaware of the gap between prosaic achievement and optimistic promise of halting biodiversity loss, but we are interested in the way in which the value attributed to mundane achievement is not troubled by the apparent lack of progress towards the more sublime goals of salvation. It seems that, for BOLI, the finite worldly pursuit of species identification takes on a different significance or quality once infused with the sense that infinite paradisiacal and redemptive futures are at stake. And as we have suggested from the various historical examples provided above, the very possibility, however remote, of therapeutic futures instils a more lofty sense of accomplishment or of mythological hope within the concrete and attainable, and brings it ever closer to the intangible nature of future promise (Fortun 2009). It is a coupling which Rabinow refers to in his study of French DNA as a 'heightened sense of tension between this-worldly activities and (somehow) transcendent stakes and values' (Rabinow 1999: 18), and it is of course common to many global life science projects (e.g. Bowker 1994b). As we explained before, Ezrahi has described a similar kind of public role for material technologies which manifestly work successfully, but whose public-demonstrative effects secure public authority, allegiance and meaning of a far more extensive kind. The former, at least, are a *sine-qua-non* for the latter. Moreover, the discursive work to make the particular demonstrations a sufficient condition of the more general authority may be relatively comfortable for many scientists, because, as Michael (1992) has described, they are quite used to seamlessly fusing 'science in particular' with 'science in general', when it comes to public communication and promotion of their work (while publics seem routinely keen, and able, to distinguish these – distinct – objects).

In a different but related sense, the relationship between *bio-literacy* and *biophilia* is also an expression of an assumed continuity between documented accomplishments and the slightly more ambitious idea of redemptive futures-to-come. Here we return again to iBOL's website to remind us of this connection:

> What would it be like to live in a bio-literate world – a world where you could know, in minutes, the name of any animal or plant – anytime, anywhere? And not just its name but everything about it – what are its habits, is it endangered, is it dangerous, should it even be there or is it an invader from somewhere else?
>
> How could we use that knowledge to protect our planet's biodiversity and promote human health and well-being?
>
> The International Barcode of Life project (iBOL), the largest biodiversity genomics initiative ever undertaken, is unlocking the door to that world by creating a digital identification system for life.
>
> (iBOL website homepage, http://ibol.org, accessed 25 June 2012)

As we have repeated, various BOLI scientists, most notably Daniel Janzen (whose vital contribution to BOLI we described in Chapter 4) have thought (and dreamt) further about the semiotic significance of reconnecting humans to nature by means of readable signatures. The following passage is a demonstration of the ostensible playfulness, imagination and creativity of BOLI signage:

> The spaceship lands. He steps out. He points it around. It says 'friendly–unfriendly; edible–poisonous; safe–dangerous; living–inanimate'. On the next sweep it says 'Quercus oleoides – Homo sapiens – Spondias mombin – Solanum nigrum – Crotalus durssus – Morpho peleides – serpentine'. This has been in my head since reading science fiction in ninth grade half a century ago. I am sure it was in the heads of Linnaeus, Alexander the Great, and Timid the Mastodont Stomper. And it has been on the wish-list of every other human confronted with the bewildering blizzard of wild biodiversity at the edge, middle and focus of a society.
>
> Imagine a world where every child's backpack, every farmer's pocket, every doctor's office and every biologist's belt has a gadget the size of a cell-phone. A free gadget. Pop off a leg, pluck a tuft of hair, pinch a piece of leaf, swat a mosquito, and stick it on a tuft of toilet tissue. One minute later the screen says *Periplaneta Americana, Canis familiaris, Quercus virginiana,* or West Nile virus in *Culex pipiens*.
>
> (Janzen 2004b: 731)

Through his own sheer enthusiasm, Janzen has the reader anticipating a neat, pocket-sized, global biosemiotic system for all of life. For all its playful language and imagery, however, such a system is imagined as unambiguous and somehow quite disconnected from the fertile and generative human–nature relationship imagined by Wilson's *biophilia*. One organism, one name, one all-encompassing knowledge. No questions asked! The 'barcorder', it is vividly imagined, would indeed enable one to know, in minutes, the name of any animal or plant, anytime, anywhere – assuming that it had already been recorded and identified morphologically then by its DNA sequence-barcode, and uploaded into the database. It would draw upon the most extensive list of species that humanity has ever compiled. The barcode is a repeatable sign or signature par excellence – all that we have said in previous chapters demonstrates these qualities. The barcode can stand for nature at a distance, for a nature disconnected from the biological material, yet connected by the global infrastructure and regimes of standards and protocols in place. Material and digit, referent and reference are condensed and at one through this amazing technology!

Janzen's and BOLI's assumption here is that DNA barcodes are drawn into the service of 'speaking for' biodiversity. As we have already noted, each barcode is a sign, or as we have decided to call it, following Derrida, a signature (1988a). As Derrida notes, the brilliance of the device of the signature is that it stands in for the signatory. But he also develops the vital point that the presence and mobility of the signature both marks and requires a complete rupture with the signatory. The very absence of the signatory is for Derrida anticipated within the presence of the signature itself, which works as a technology in which absence is taken for granted. In the case of BOLI, the DNA barcode is mobilized and – with some infrastructural help – speaks for species in the absence of those organisms as signatories that have yielded their DNA for the fabrication of a barcode as signature. At first glance biodiversity is intrinsically *present* in the barcode. But in another way, the potential for DNA barcodes to speak for organisms takes for granted their absence, just as BOLD anticipates loss of diversity (Chapter 6).

One of the potentials for 'distortion' in the semiotic system we are accounting for here then is the risk of *forgetting* that a signature can and indeed should operate independently of the signatory, even if ultimately the latter can be seen as the source of the former. As Derrida suggests, a sometimes assumed sense of continuity or homogeneity between the signature and the signatory has the effect of investing a distorted degree of presence of the signatory in the signature itself, forgetting the 'needs' for a quite different quality of presence of the signatory.[9] As we read Janzen's fantastic portrayal of the possible, we recall here that DNA barcoding related websites talk only of species lists and the importance of naming – or to put it another way, of the importance of bio-semiotic signatures themselves. They do not convey the indeterminacy of the relationship between signature and signatory. They do not evidence concern at the one-time and possible fragile presence of the signatory (a single confirmation that a voucher specimen is allowed to stand in for the whole of a species). They do not point the reader to consider the heterogeneous relations required to be in place to ensure the signature or signatory's continuing viability.

This prompts a concern: that the immortality of barcodes, housed in a smooth operating data system, may blind BOLI to the fragile mortality of material biological species and organisms themselves. The reification of barcodes may paradoxically encourage a forgetting of a more expansive and contingent world of biodiversity within which all the complex, different, emotional and sensate human connections and experiences will hopefully perform *biophilic* therapeutic futures. In other words these acts of reduction in which the fabrication of the barcode takes on sacred proportions could work to anaesthetize feeling and care: the agony of biodiversity loss (past and future), as well as the ecstasy of sublime reconnected and flourishing futures to come. Furthermore, by failing to acknowledge and work with the indeterminacy of the relationship between signature and signatory, barcoding marks are incapable of managing the unknown and heterogeneous relations required to be in place to ensure the signature's continuing viability, in a larger milieu of changing biodiversity and its changing stressors.

In this section above we have conveyed the different ways in which BOLI holds closely together a range of mundane achievements and more profligate promises of transformative futures-to-come. And we see the DNA barcode as a signifying device, potentially capable of linking the prosaic and the intangible at the centre of these efforts. We suggest in this way that BOLI manages, with considerable dexterity, to connect promises (of a future that may never see fulfilment) to more mundane practical work in the present. What interests us in particular here is the work done for BOLI by these gaps, but we are also troubled by the sense that BOLI may be short-changing itself by failing (so far) to develop a sharper sensibility to the limitations of the signature (DNA barcode), thereby denying the open-ended nature of human relationships and meanings with(in) biodiversity and possible futures to come – all phenomena which, as we have demonstrated, are latently communicated by the DNA barcode.

The promise

In order to bring further insight to our understanding of what BOLI gains by moving between and implicitly connecting (in tension) quite different registers of attention and possibility, we turn finally to Fortun's spirited but cautionary exploration of the creative dynamics of promissory science and speculative finance in the world of Icelandic Genomics. His analysis of the attempt by the American company DECODE (founded and led by a US-resident Harvard University Icelandic geneticist) to set up a DNA database derived from the blood samples of the entire Icelandic population, draws extensively on the idea of paradox, or multiple paradoxes, which bear considerable resemblance to those we have explored so far in this chapter. Paradoxical couplets such as achievable/promised, innocence/ownership; possible/impossible, transcendental/pragmatic; true/false; trustworthy/shady; sincere/insincere, work, in his text, as organizing principles not only for genomic science itself, but for his own writing. These are qualities which for Fortun are 'conventionally distinct' but also 'mutually co-dependent' (2008: 13). Like the Icelandic landscape, they are entangled and fissured. The rifts, or chiasmas, between terms, suggests Fortun, consist of hopes, dreams, fantasies and material conditions which cannot make sense together and yet cannot make sense apart.

Ultimately, however, it is the spectral quality of promissory futures-to-come which endows these couplings with an air of productive indeterminacy. This was indicated in our earlier biosemiotics discussion using Derrida, of the ambiguous coupling between signature and signatory. The fissure, or chiasma, is an image which helps Fortun understand the nature of technoscientific promise as both rooted and untethered. But what is most important here is the fact that the promise of the future takes on *more meaning* than what is actually achieved by science in the present – and what will materialize in the future present will anyway be different. Potential therapies-to-come excite more and galvanize more financial and public buy-in than therapies already on the market. Drawing upon Austin's speech–act theory and Derrida's deconstructive approach to promise, Fortun emphasizes how promise by definition cannot coincide with itself, it always produces an *excess of itself*. And he speaks here, in a way relevant for our analysis, of the '*artful confusion*' of keeping in view the quite discontinuous actualities of pragmatic achievement and future possibility (or past 'future possibilities' now-presently accountable – even if not put to the test). This is a confusion, or perhaps even an act of deception, which implies that the texture or reality of promised futures cannot be firmly accounted for. Focus lies instead on the tangibility of pragmatic achievements, and eyes turn away from the fact that these rarely if ever coincide with the promised future itself as it 'becomes present'.

Fortun realizes that we could be tempted to interpret promissory acts be they in science or beyond as somewhat deceitful or reckless. Yet, in a chiasmic sense and faithful to his analytical commitment to indeterminacy, he chooses in part to celebrate the playfulness derived from the fact that promised futures are unavoidably in excess of, and cannot fit, the present. He calls upon his readers to observe as creative and dextrous the scientific ability to portray and thrive on the energy generated by such a paradox. However, he simultaneously takes on a more cautionary tone as he pursues further the idea, derived from Austin's work on the *Scandal of the Speaking Body* that speech acts of promise ultimately create free-floating promises that become separated from an intentional subject. As such those uttering the promise cannot be held accountable to the future but only to present achievement, 'promising, after Austin, does not require intentionality; saying will do' (2007: 106). The utterance of the promise, in other words, often does the 'job in hand' – reassuring an anxious fellow-being: calming a mood; sealing a deal; establishing a collective commitment.

For BOLI, indeterminate and chiasmic couplets of the kind identified by Fortun work in two ways. Such couplets hold spaces between fears of devastation and hopes for redemption on the one hand, and between the practical work of gathering species information and the possibility of ecological and human therapeutic futures on the other. Mythological hope tethers these couplings to the achievable but also allows radical untethering (because future redemption by definition cannot be defined). Both futures act as a rich resource to secure investment in BOLI and propel further endeavours in scientific research and public enrolment alike. Thus, BOLI requires mythological paradox as vital resource.

Conclusions

In this chapter we have been fascinated by the ability of BOLI scientists, as part of their repertoire of the arts and skills of prediction relating to species presence and loss, to hold in tension expansive claims for human and planetary redemption and much more bounded claims to scientific precision and the banking of species knowledge. We have been alert therefore to the complex meanings attributed to species loss. In so doing we note how human and cosmic frailty are and always have been a vital resource for technoscientific research. On the one hand, by locating BOLI's performance as a co-dependency between the dual promises of the mundane and the sublime, a performance consistent with several centuries of entanglement between theology and technoscience, we note that there is nothing particularly extraordinary here. Scientific promises of deliverance for humankind are not new, and genomics has become almost definitive of this precarious ambition since its launch in the late 1980s.

We have stressed that BOLI's injecting of transcendental other worldly practices (a world-to-come whether paradisiacal or dystopic) into those that are very much located in this world, is quite an achievement. It is an achievement which demonstrates how vivid scientific imaginations play a vital role in BOLI's beguiling optimism. On the other hand, we have explored here how these grand promises sit uneasily with natural and healthily sceptical questions about how and when these promises can be tested in observable empirical worlds – worlds where the work reflecting their persuasive success is thick with frictions which will need to be arduously accommodated, lubricated and overcome. However, the common – and on the face of it, justified – complaint that the promised land of barcoding nirvana is like the human genome or stem cells equivalents, forever 'just around the corner', is perhaps missing the point. It has been noted before that belief in a reality that is 'just around the corner' is actually a corollary of the dogmatic state of much normal scientific culture (Kuhn 1962), and that scientific expectation of predicted effects can sometimes be so deep as to become as real as actually observed ones. Cases have been documented where this syndrome has appeared to prevail, without evidence of deliberate deceit (Keepin and Wynne 1984; Nightingale and Martin 2004). In other words, these redemptive promises in genomics at large, as in barcoding, are presented as propositional claims about how the future will be, just as normal scientific knowledge offers (in principle testable, even falsifiable) propositional statements about the present. But the apparent ease with which any questions as to why past scientific promises, along with the constant production of new ones, have not been met/answered, encourages a different interpretation of this curious relationship between redemptive promises and mundane and ordinary outcomes embodied in the same rational scientific programme.

We have thus taken a more cautionary angle in this chapter, by asking whether BOLI may be forestalling its own emancipatory potential as a technoscientific venture with transcendent hopes for the future. We suggest that BOLI exudes a certain complacency in accepting that the DNA barcode works 'for now' as a valid and precise signature for life's diversity; the DNA barcode becomes in other words a reification in the minds, practices and commitments of BOLI scientists. In accepting this, BOLI risks, we argue, a serious forgetting of the subtle, complex and heterogeneous relationships across the world between complex human subjectivities and the natural world, relationships upon which the DNA barcode depends if it is to fulfil its promises not only of accurate species documentation but also of enriched human–natural biophilic reconnection.

We therefore end this chapter by focusing upon the scale and the duality of the responsibility shouldered by BOLI scientists as they hope for and promise futures of both bounded and boundless proportions. If they commit to the role they have apparently donned – as prophets for nature and society – there is, we argue here, much more to do. We are thinking here of a vital need for BOLI not to allow itself to be seduced by the revered status it affords to precision – by drawing, as it does, the 'sacred' into the 'concrete' – but to pledge to nurturing their own (and their publics') imagination concerning both the remit and the limitations of the DNA barcode. The DNA barcode, as the quintessential signing device we have described it to be, stands not only for species diversity but also for the human cognitive and sensory relationships invested in engendering a more intimate and caring set of connections between humans and the organisms with which we co-habit the planet. The imagination we are alluding to here is one which would conjure a real sensibility – with both practical and imaginary/poetic dimensions – to what it might mean to 'live with' (Haraway 2008) the diversity which we and the planet require as we look and live towards the future.

8 What is it?

Identifying nature and valuing utility

> When naturalists introduced the word 'biodiversity', they had no idea that a few decades later they would have to add the proliferation of surprising connections among organisms the proliferation of many more surprising connections between political institutions devoted to the protection of this or that organism. While naturalists could previously limit themselves, for instance, to situating the red tuna in the great chain of predators and prey, they now have to add to this ecosystem Japanese consumers, activists, and even president Sarkozy, who had promised to protect the fish before retreating once again when confronted with the Mediterranean fishing fleet.
>
> (Latour 2010: 480, *New Literary History*, The Johns Hopkins University Press)

Musing on the shifting interconnections of our world and its knowledges, Latour notes the recent growth of naturalists' 'networks' (or 'connections') to incorporate the affiliations of consumers, fishermen and other nature interveners, activists and politicians, all of whose multifarious intersections with the bare biological species, the red tuna, have both grown, and become more evident, as its political economy, along with ecological concerns, have gone global. The red tuna itself, he suggests, has also extended its own ecological territory to take in sushi bars planet-wide! The complex interconnected world that Latour describes (swarming with the agencies of humans and our nonhuman cohabitants) is a world instantly recognizable to many barcoders and barcoding users, many of them also 'naturalists' or even environmentalists of a kind. But more relevant to our present concern, this is a world in which the role of science ('pure' and 'applied'), is itself shifting in relation to politics and governance. Latour and STS at large has been asserting for some time that we are no longer in the kind of universe where Nature can be studied by disinterested scientists to help determine what to do (a politics-based on the facts of the matter) (Latour 1993, 2004, 2010).[1] We are, instead, in a messy 'pluriverse' in which non-human destiny (nature) and human destiny (politics) are entangled for everyone to see. Of course this is also the entangled world which many barcoders now take for granted, and in which they feel that barcoding science *has* to work, *has* to be useful, and *has* to show its relevance. As we noted earlier, especially in Chapters 4 and 5, BOLI recognizes and pursues wholeheartedly its need to network and entangle globally.

It achieves this with multifarious other actors and their own messy networks, and keeps its own discursive purifications more-or-less intact, even across such dense and extensively distributed centrifugal entanglements, by its strict adherence to the virtues and disciplines of its technical standards and the full analytical chain (from diverse 'wild nature' field sites, to lab and digital translations, to database, and back again). This allows a purified Nature which transcends the entanglements to be reasserted, even as those entanglements are still being extended in practice. Under these conditions, we also note as relevant for this chapter, that the technical practices, choices, commitments and questions which make up barcoding science, are imbued with (anticipatory, as well as immediate) imaginaries of those distributed actors and their situations. This increasingly includes possible users, uses and possible user-relationships, where the distinction between 'user' and 'scientific collaborator' can never be entirely clear.

In this chapter we connect Latour's portrayal of the pluriverse, in which science, nature and politics connect in fecund new relationships, with contemplation of the much hailed usefulness of barcoding. As we saw in the previous chapter, BOLI and barcoding advocates have become skilful over the past 10 years in creating, and sustaining, grand and even sublime claims for their technoscientific enterprise. But as we have also noted in previous chapters, the very first papers emanating from Guelph articulating these boundless promissory claims always also included much more mundane ideas about the potential uses, and usefulness to society, of barcoding. The overt and covert achievements and meanings of these co-existent claims are dealt with here. But from the very inception of barcoding efforts, such claims-making was also partly skilful positioning. Taxonomy, writ large, had been perceived to be a discipline that was under-utilized, stuck in an 'ivory tower', too arcane, remote and complex to be of service in most areas of public policy, even its most immediate 'obvious' user-field, biodiversity policy. Barcoding, in deliberate contrast, was portrayed as highly useful, 'out there', versatile, in touch with policy, and able to serve a range of needs in society. A focus on the utility of barcoding was part of what made this new version of taxonomy sparkle in the eyes of potential funders: as emphasized in Chapter 2, its *value as a tool* was constantly impressed upon audiences and on the readers of a steady stream of journal articles advocating a barcoding approach.

An emphasis on utility, of course, does not look at all surprising in the wider science-society context of the early 2000s, indeed it has been essential to the survival of many disciplines. In the late twentieth and early twenty-first centuries, two contrasting moves have combined to underpin a steady shift in our understandings of knowledge production for society. The first has been a relative increase in *private* funding for scientific work conducted in publically funded institutions like universities, museums and research institutes (Callon 2004; Busch 2007; Etzkowitz and Leydesdorff 1997; Webster 1989; Stengel *et al.* 2009; Mirowski and Mirjam-Sent 2002). The second is a simultaneous requirement upon researchers to demonstrate the *public or private utility* of their work (a utility currently measured, in the UK at least, as 'impact'). These two convergent movements have not only caused seismic shifts in the private/public boundaries of knowledge production and research, they have also put a premium on use, utility and social relevance (Strathern 2004), including for what was once thought of as academic (equals pure, or basic) science. To put it another way, the values supported by private funding of research (project-based, discipline-indifferent, timely, economic, targeted research) and the values associated with public good (use, utility) have become twinned in new ways. The 'pluriverse', and barcoding within it, we suggest, are a product of these recent reconfigurations and their tumultuous public/private re-orderings, co-productions and borrowings.[2] This eruption of an ambiguity which has always existed in the self-definition of 'the purity of science' (Edge 1975) is also reflected in the ambiguity of BOLI's self-definition as pure/basic research, for example to demarcate its distance from commercial science in its negotiations with the CBD for access to crucial materials from politically contentious source-regions, as described in Chapters 4 and 5. Against some yardsticks as to whether it has commercial purposes or not, this is appropriate; but against others, it is clearly an applied or at least mission-oriented science, with a prominent mission for the world. Co-production, in particular, is important here – the new pluriversal networks that Latour and others describe could not have been made and cannot continue to be made unless a 'public good' is imagined to go with it.[3] This requires cutting existing networks as well as making up new ones, as we shall see. But from this cutting and creating, this severance and composition, an imagined world – a pluriverse including barcoding and new human–nature orderings – is being crafted through a dominant emphasis on imagined utility.

The value of use

As we have briefly described in Chapter 1, at an early point in our research back in 2006, we rehearsed what we had learnt about the many amazing attributes of barcoding with a group of taxonomists at the London Natural History Museum. We were told, in so many words, not to be so gullible as to believe all the fantastic claims we had heard and read about, and instead to look upon barcoding as 'only a diagnostic tool'. Barcoding, it was asserted would *not* re-order existing ways of knowing the natural world, nor would it significantly re-order taxonomy, nor our relationship with nature; and it could have little relevance for the wider understanding of organisms, species, and their development or evolution. It would only offer a new technology for *the identification of known species*, but, as such, it would be a valuable tool. It would be particularly useful where morphological methods were weak: in the identification of cryptic and polymorphic species; for species where morphological observation is extremely demanding; and where the means was needed to associate (morphologically different) life-history stages of unknown identity. Barcoding would also be helpful in cases where morphology is ambiguous or uninformative, or where only part of an organism is present for identification purposes. When organisms presented for identification had some positive or negative implication for economic, regulatory, clinical or other applied reasons, the value of their swift, precise and easy identification would be immediately apparent.

After some confusion, we interpreted these deliberately reductionist descriptions of the technical attributes of barcoding as one way of ignoring and thereby disenfranchising the grander claims being made about barcoding. Many taxonomists did not want to buy into the idea that barcoding constituted a viable universal standard theory-method for twenty-first-century taxonomy. They saw barcoding as parasitical on the long and extensive legacy of organized taxonomic labour and at risk of unravelling and usurping this legacy with a more superficial public as well as esoteric scientific understanding of nature. On the other hand they were prepared to accept the benefits of standardizing barcoding methods for performing user-friendly, intellectually undemanding identification tasks for which non-taxonomy users might be willing to pay. But, even so, critical questions around the value of such use and utility were in the air, and still persist as we write. The boundary between convenient and limited short-circuit use-value, and superficial misrepresentation of nature, is inherently ambiguous, and more generally so across science than only for taxonomy. Some taxonomists felt, and still feel, that barcoding's orientation towards utility, in effect portrayed taxonomy as a mere 'service industry', not as an independent, research-driven science. This, it was feared, might have the effect of dumbing-down the intellectual basis of taxonomy just when it was striving, with some success, to escape its dismissive representation as 'stamp-collecting' not science, and in ways which might potentially damage its wider intellectual goals (Will *et al.* 2005; Wheeler 2008; Ebach 2011; Ebach and de Carvalho 2010; Wheeler *et al.* 2012). Ebach and de Carvalho still warn that an 'unreflective instrumentalism' may be populist and attractive to funders, but cannot support what they suggest to be a 'valid' more basic scientific enterprise of on-going, diverse, integrated – and often necessarily slow – taxonomy (Ebach and de Carvalho 2010; Wheeler *et al.* 2012). While developing and promoting its utility, globally, BOLI's leading scientists are aware of these deeper issues, and carry a faithful sense that their black-boxed and automated means for users and publics to encounter nature will eventually generate new human and even natural ontological conditions which will offer us the knowledge to engender mass *biophilia* and to protect the diversity and health of the biosphere. We come back to this optimistic vision towards the end of the chapter.

For now we note that in the 10 years since Hebert *et al.*'s first papers started appearing in the taxonomic literature (2003), barcoding has indeed started to demonstrate its practical utility in a whole host of applied and policy domains, including health, agriculture, environmental monitoring and various specific kinds of regulation – as any glance at the steady flow of publications being posted on the CBOL and iBOL websites will indicate. To give the reader a sense of what this utility amounts to in practice, we provide three diverse illustrative examples of the use of barcoding – in the identification of fish catches for consumer and fish protection, in the detection of invasive carp species in the waters of the North American Great Lakes, and in the identification of African mammal DNA in the guts of the tsetse fly.

'What is it?' fish forensics as public service

An investigation by the *Boston Globe* daily newspaper reporting on the widespread mislabelling of seafood in Massachusetts in October 2011 seems to exemplify the 'pluriverse' in which barcoding is trying to find its place. Working together with Smithsonian Museum taxonomists as they piloted a new barcoding protocol for amplifying and barcoding fish DNA, *Globe* journalists revealed that 48 per cent of the fish that their reporters purchased in restaurants, grocery stores and markets had been mislabelled. Increasingly depleted (and so, expensive) sole and red snapper, in particular, were being surreptitiously substituted with less expensive types of fish. For the *Boston Globe*'s readership, the message conveyed was that the public were being ripped-off and their consumer rights abused! Identification of fish species, once fish are prepared for sale or consumption, is difficult even for an expert eye. Most fish consumers would have little clue that they were not buying what was stated on the label, or not eating what was listed on the menu. So here, reported the *Globe*, is where fish sourcing and fish sales meet the power of CO1 for detecting difference between fish species (Ward *et al.* 2005). A further article in the popular science media reports that barcoding can also pinpoint the circulation in markets and restaurants of endangered species. Some species of bluefin tuna have been fished to near extinction, according to scientists at the Smithsonian, and many sushi restaurants still serve those species.[4] The logical first step is to identify it. Smithsonian taxonomists believe that consumers will soon be able to use a handheld barcoding device to identify 'any number of the hundreds of rare animals that might wind up on their table or in their shopping bag'.[5] As Lee Weight, Director of the Smithsonian's Laboratory of Analytical Biology put it, the technology that will enable this vigilance is not that far off. He imagines that users will soon be able to 'rub a smart phone against a sample of fish and it will connect to the fish DNA barcode database, and identify a fish as red snapper or what have you'.[6]

The report in the *Boston Globe* and the story of tuna detection have been taken up and celebrated as success stories within the Smithsonian Museum, Washington – the home of CBOL, and FISH-BOL (a global Fish Barcoding of Life Initiative).[7] The Smithsonian and their partners had good reason to celebrate. Both stories demonstrate that barcoding is indeed being *used* in the *public domain* for a very *practical purpose* – to detect and possibly prevent the mislabelling of seafood, thus to clean up not just fish marketing, but fishing and marketing upstream in the whole supply chain. Through routine detection of this kind of deception in the United States, it was hoped that fish market racketeers would, in time, avoid the US market altogether. This is taxonomy with teeth! 'Taxonomy is not dusty old museum science the way it was when I first got into it', reflected Lee Weight.

Is it there? Detecting invasive species

Our second example of barcoding being put to practical use looks at what happens when the question, 'what is it?', requiring a piece of biological material in the laboratory to be identified, changes to the question, 'is "it" there?' We refer here to the detection of unwanted invasive species in the vast waters of the North American Great Lakes.

Darling and Mahon (2011) have written an illuminating account of the utility of DNA-based methods to detect aquatic invasive species (AIS). One example they give concerns the importing of cultured oysters from Denmark to the Netherlands.[8] As they relate, in the spring of 2010, microscopic inspection of these imported oysters revealed the presence of egg capsules morphologically similar to those of the invasive Japanese oyster *Ocinebrellus inornatus*. Standard barcoding methods, involving the amplification and sequencing of CO1, were used to ascertain the true identity of these hard to identify egg capsules. When the results indicated unambiguous matches to *O. inornatus* reference sequences in BOLD, the oyster shipment was quickly quarantined. The management benefits of this use of barcoding techniques in the timely detection of suspected invasive species were clearly spelt out:

> Not only do DNA-based methods in these cases provide a level of certainty in specimen identification difficult to achieve through traditional morphological approaches, the time savings afforded by molecular confirmations can be critical to effective management. The possibility of confirming AIS detection in hours to days instead of weeks to months allows managers to act quickly, thus minimizing the risk of AIS spread.
>
> (Darling and Mahon 2011: 979)

But Darling and Mahon are also interested in the *extension* of such DNA-based methods for the detection of AIS *propagules* in environmental samples (e.g. lake sediment cores, ballast water samples, etc.). By propagules they mean a part of the organism that becomes detached from the rest, but has the capacity to form a new organism – a piece of regenerative plant material, an egg, or similar. What they suggest here is that the technical challenges associated with detecting target species in complex watery samples are significantly greater than those associated with individual identification of specimens via barcoding (Darling and Blum 2007).

The experience which they have of this derives from the use of DNA-based monitoring for invasive Asian carp in the Great Lakes region of North America. As they explain, Asian carp were introduced into North America in the 1970s as a measure for cleaning aquaculture facilities, but they escaped from containment ponds in Arkansas into the Mississippi River basin shortly thereafter (Darling and Mahon 2011: 980). Upon release into the wild, two species of carp spread northward, and by the mid-1990s were highly abundant in the Illinois River, which links directly to the Great Lakes via the Chicago Sanitary and Shipping Canal (CSSC). Their presence in the Great Lakes had the potential to devastate the natural aquatic fauna of this system, it was feared. Given known limitations of traditional detection methods (primarily netting and electrofishing) at low population densities, standard molecular techniques (DNA extraction, species specific PCR, electrophoresis) were used to detect target species DNA present in material such as skin cells, mucus secretions and faeces released into the aquatic environment. If the target carp DNA was detected, this would allow *inference* of carp species presence. When the results of these studies found DNA from both carp species throughout the CSSC, and in one case as far north as Lake Michigan, the alarm was immediately raised. A number of groups argued for full separation of the Great Lakes and the Mississippi River basin and pushed for closure of the hydrological locks that lead into Lake Michigan. This was met with strong and vocal resistance from local waterway operators who routinely use the canal and a State-led, court-level controversy about the connection and separation of these water bodies ensued. We pick up and complicate some of the interesting implications of this story later on in this chapter, but for now, we move on to our third example of barcoding's utility.

What is there? Barcoding and public health

Our third example explores what happens when barcoding techniques are applied to answer a very open question – 'what is there?' In the example we give, what is being sought are traces of DNA from the blood of a wide variety of African animals that might be found in the gut of a tsetse fly – almost an open-ended search 'in the dark'. Here, the value of barcoding is being positioned to play directly into urgent matters of public health, since tsetse fly are carriers for the widespread and devastating sleeping sickness infective agent.

We refer to an entomological study carried out in Kenya, Tanzania and Uganda where scientists, one of whom we met at an African regional 'leading labs' meeting on our global CBOL travels, have collaborated on a project to track the feeding patterns of Tsetse flies (*Glossina* genus) (Muturi *et al.* 2011). They have done this tracking by analysing the 'bloodmeals' of the Tsetse using barcoding techniques (finding and analysing CO1 genes to detect the species of animal that gave its blood to the tsetse fly). As in the two other applied studies reported above, these scientists emphasize the urgent social need to carry out this research: in this case, scientists are keen to understand the life habits of the tsetse fly; to identify its food sources; to understand better the way it transmits its deadly and destructive disease; and to find ways of trapping and eliminating the fly.

Muturi *et al.* (2011) report that, following large epidemics of sleeping sickness at the beginning of the 2000s whole village communities in Central Africa have been wiped out. Unacceptable levels of sleeping sickness continue to be reported in several areas and livestock are constantly under threat of infection. Animal sleeping sickness (*nagana*) in Africa has largely excluded livestock from approximately 70 per cent of the humid and semi-humid zones of sub-Sahara and about three million cattle deaths have been suffered. With cattle deaths, of course, comes livelihood-devastation, and human poverty. Estimates of losses to Africa's Gross Domestic Product (GDP) are in the region of 4.5 billion US dollars per year.

As these authors explain, it is notoriously difficult to trace what food sources tsetse flies obtain, and yet information on the source of blood meals of tsetse fly vectors is essential in understanding the relationship between hosts and vectors, and their respective roles in the sleeping sickness transmission cycle. Since different species of tsetse seem to feed from species of cattle and wildlife, knowledge of their food source can help in the development of odour-attractants that are used in tsetse fly traps loaded with insecticide to kill the flies. The scientists dissected the 'midguts' of a sample of tsetse flies from across different regions, extracted the blood, and amplified the COI gene from each sample. COI analysis using BOLD's identification engine revealed that, in one region sampled, six tsetse had fed on African savannah elephant, five on warthogs, one on African buffalo and one on baboon! In a different geographical region, all 12 flies had fed on cattle.

Barcoding is advocated as having a highly practical role in the design of public health strategies here. Knowing the preferred hosts of tsetse bloodmeals can provide a more targeted approach to creating olfactory cues, which in turn can improve the efficiency of tsetse traps. Additionally, if tsetse in one region are found to be feeding exclusively on the blood of cattle, insecticide can be used on the cattle themselves as mobile traps (Muturi *et al.* 2011).

Complicating stories of use and the pluriverse

We must now pause to consider how to think about these stories, which so clearly state the public good associated with the use of barcoding. It is hard, given the three narratives presented above to see any reason to be concerned about barcoding's identity as a useful science, releasing value through the intersections between nature, science, politics and human needs in the messy pluriverse. But, in truth, we have truncated these stories: we have conveyed above only half of what each one of them has to tell, and we pick up some important threads from each one of them in the following section. We have done this to show that these stories can be conveyed as unalloyed successes. But as many authors reflecting on the rapid transformations of knowledge production towards use, utility and the public good have suggested, success stories like those portrayed above, often obscure more complex underlying reconfigurations that deserve equal attention, and more searching questions. The dynamics of the pluriverse require more focused analysis and we first develop a sense of what it is we need to look out for, before we return to complete each of the three narratives.

An initial set of preoccupations bubbling up under the transforming circumstances of knowledge production in the 1990s and early 2000s, for example, is about recognition of the quality of the epistemic contribution that applied knowledges can bring. In their by now well-known characterizations of Mode 1 and Mode 2 knowledge production, Gibbons *et al.* (1994) suggest that, in Mode 1, science enjoyed societal legitimacy as a distinct source of privileged knowledge. Science was produced and validated in separate arenas (such as laboratories, and the scientific community with its distinct normative form) before being shared with society, and scientists saw it as part of their task to use this knowledge for the benign transformation of society, according to 'modern' principles. In Mode 2 science, on the other hand, this external position and authority has eroded (see also Ezrahi 1990), with many more social actors entering into processes of knowledge-production, including its previously distinct processes of validation and correction. Consequently, under (current) Mode 2 conditions, science, as a (less clearly defined) 'internal' set of practices has to create links laterally, into society, and these links and connections, whilst contributing knowledge, cannot necessarily bring definitive closure to applied knowledge issues, problems or controversies. In contrast to the classical understanding of Mode 1, the knowledge-problem agendas for Mode 2 are not set only by the scientific research community, but also by a range of societal stakeholders. Whilst bringing fresh knowledge to bear, Mode 2 knowledges also introduce deeper elements of uncertainty, ambiguity (what is the problem here?), instability and contingency to knowledge making. The authors' point is not that such uncertainties can be resolved or erased through better science, but that their inevitability and thus normality requires explicit recognition and attention (Gibbons *et al.* 1994; Nowotny *et al.* 2001: 2). We will see in the next section the relevance of this point for the detection of carp in the Great Lakes.

A second concern has been how to sustain mutually productive relationships between more applied circuits of knowledge, on the one hand, and their roots in 'disciplinary' modes of knowledge production, on the other (Strathern 2004). Marilyn Strathern referred to the whole shifting scaffold of knowledge production in the early 2000s as the 'contemporary intensification of debate over the relationship between knowledge and the public good, and how creativity can be pressed into public use' (Strathern 2004: 45). Interrogating the implications of the creation of a new knowledge *borderlands* – sometimes seen as straightforward commercial–academic transactions, but more often subtle combinations of 'interdisciplinary', 'applied', 'translational' or 'engaged' science, as in Mode 2 science – she points out that it is not only the commodification of knowledge or issues of ownership, or intellectual property that are at issue in this new terrain where the utility and application of knowledge is at a premium. Questions about new *flows* of knowledge quickly come to the fore. These concern the kinds of relations that need to be in place to get such flows to make reciprocal loops and therefore to have a corresponding *moral* circuitry which is needed for all social organization, including for knowledge-producing organizations like science (Shapin 1994).

We shall see below that this point relates to the forensic determination of fish species in markets and restaurants, but we look in more detail first at what Strathern means. Reflecting upon a question sent to her from a knowledge expert from her own 'field' of research in Papua New Guinea, Strathern explores the issue of responsibility in the making of knowledge:

> When a woman goes in marriage to another clan, and bears many children for that clan, will her in-laws think on the kin from which she came? When someone plants a seedling, and that seedling grows into a tree, and the tree bears much fruit, will the eater of the fruit think on the person who planted the seedling? 'Thinking on' is an orientation to another person that is meant to summon all the strands of obligation and mutual acknowledgement/

recompense that put people into one-another's debt … The question he was putting in front of me was the question of responsibility, of moral authorship if you like. The question was of equitable reward. Not a determination of who has the rights, but of how we were going to conduct our relationship.

(Strathern 2004: 61)

Strathern complicates the questions about the future use of knowledge, as in the future fruits of the planted seedling, questions which are intrinsically open, and in need of continuing responsible address. As we continue our stories about barcoding's uses below, Strathern's interpretation – that the questions put to her concerned the continuing, not one-off, question of responsibility, moral authorship, and the open-ended challenge 'how are we going to conduct the relationship(s)?' – captures precisely the sentiment driving the debates of many barcoders as they narrate their new practices in the practical application of barcoding. Their preoccupation, to put it clearly, is how to conduct a recipro-cal relationship between a *useful* barcoding – focused on the identification of known species, or perhaps even on species-indifferent diversity-counting as a measure relevant to a monitored habitat (more on this below) – and the dis-cipline of taxonomy. This open-ended question, for barcoding and taxonomy, concerns the issue of epistemic responsibility for creating and maintaining understanding about species, speciation, and biodiversity as well as about 'what things are' (slabs of fish in markets and restaurants), 'whether they are there' ('invasive' carp in the Great Lakes) or how organisms thrive and might most easily be eradicated (tsetse flies in the African bush).

The question of a kind of epistemic reciprocation between barcdoing and taxonomy is however not all that concerns barcoders in the present time, where knowledge is being 'enterprised up' (Strathern 1992; Haraway 1997) and where 'creativity is being pressed into public use' (Strathern 2004: 45). Indeed, in the last section of this chapter where we discuss the use of barcoding in environmental monitoring we suggest that the question of reciprocity is overtaken by a creative drive by or through science, to redefine what *is* publically useful, thus to define what is the (proper) public good. We show here that the value released from bar-coding is 'on the move'. Using barcoding brings new circulations of knowledge and value into play. Here, taxonomic knowledge is not only 'enterprised up' but 'enterprised out' (Hayden 2003a: 43). It comes from somewhere, but it generates new value elsewhere. It is oriented outwards, towards new needs and develop-ments in emergent natural–social relations which have their own dynamics and circulations. Thus a value is derived from barcoding's use that depends upon taxonomy but isn't necessarily aligned back into the disciplining that taxonomy requires and promotes. Instead it is generating new, outwardly proliferating con-nections and relations that are potentially transformative for taxonomy and are pushing the imaginations of bioinformaticians to detect new and different ways of perceiving, perhaps (re)classifying, and protecting nature. We return now to repair our truncated stories.

Fish forensics and reciprocal ties

We now pick up the story told in the first half of this chapter about fish forensics, recalling that the *Boston Globe* newspaper had published an article in 2011 celebrating the achievements of the Smithsonian Museum in barcoding fish flesh in markets, thereby detecting mislabelling of expensive and/or endangered fish species. What we have not so far documented is that, in the *Boston Globe* article, it was clearly important to the Smithsonian to demonstrate its concern to nurture relations between the application of barcoding, on the one hand, and taxonomy on the other. A moral circuitry, a reciprocation back into the disciplining practices of taxonomy was deliberately included in the *Boston Globe* report. Lee Weight stressed clearly that the simple question asked of a fillet of fish – 'what is it?' – required not only DNA sequencing and a system of referral to a reliable reference library in BOLD, but it also required *good taxonomy,* not only good as science, but also capable – a more stringent discipline – of upholding the law (see also Hanner *et al.* 2011; de Carvalho *et al.* 2011).

> Where fish are concerned, no one can rival the Smithsonian in our long-term collections, professional taxonomists and our DNA laboratory experience. We have a DNA database of barcodes that we developed for hundreds of species of fish. Any laboratory can extract and sequence the barcode for a species of fish, but each of our barcodes is backed-up by a real specimen in the Museum of Natural History collections which was first identified as to species from its physical characteristics by one of our expert taxonomists in the Division of Fishes.
>
> (Lee Weight in interview with the *Boston Globe*)[9]

Thus, the argument is made that applied barcoding and pure taxonomy ('dusty old museum science') are shown to need each other and to be mutually productive. In order to be useful barcoding needs organized collections, specimens, taxonomic work-legacies in the form of records, data-bases, voucher-specimens, and much else, as well as trained taxonomists who can first identify the species through its morphology. The pluriverse, in Latour's terms, may involve new and messy entanglements between nature and politics, but it also involves a politics: even the most go-ahead 'Mode 2' barcoders are aware of this politics, subtly arguing for a reciprocation from barcoding to taxonomy, especially of the 'traditional, morphological' kind! Barcoding needs all of this for its own proclaimed usefulness, and for demonstration of its applied value. What this story also highlights, therefore, in addition to the success of CO1 identification in an applied setting, is acknowledgement of the dependency of barcoding upon established and well-funded modes of taxonomic knowledge production, underpinned by materials, skills, training and collective dedication. This is a virtually hidden investment resource which barcoding mines, every time it performs barcoding's 'public good'.

Invasive species detection involving uncertainty

Recalling now our second example of barcoding use in the detection of invasive carp species in the Great Lakes of North America, we have not so far narrated the warnings that Darling and Mahon offer to their fellow scientists and barcoders about the complications that the use of barcoding techniques in such highly public and political contexts can bring about. In this case, micro and macro apparatuses for detecting invasive species make for some uncomfortable incommensurabilities in the full public glare. Darling and Mahon stress in their paper that a number of false positives can be created through these sensitive methods of detection. Carp may be 'detected' without being present, having left their DNA 'signatures-now-barcodes' behind. The greater sensitivity of DNA-based detection methods of course can create evidence, but it can also create greater interpretive uncertainty and contingency into the processes of knowledge production intended to clear up an ambiguous matter of whether particular carp species are there, or not. Greater sensitivity can equal 'false positive' detection, with ensuing error-costs which could be very considerable. Sensitive methods can also create a further need – to bring into the 'pluriversal, Mode 2 borderlands', more contextual biological and ecological knowledge of fish and preying bird habits, and a range of other factors which might affect 'collateral' DNA distributions with no fish present.

As an aside here, but one that we feel may be relevant in this kind of case, the danger which we discussed in Chapter 7, that the signatory (here, the carp species at issue) might be fused with the signature (the DNA barcode) would seem to be non-existent, since it is just this relationship, and whether it is an authentic or a spurious presence in a particular time and space, which is up for investigation. However, we note that the distance or 'presence' which Derrida emphasized is a function of richness or poverty in the interpretive sense. Our point there was more about the capacity of barcoding technoscience to maintain a poetic distance (Derrida would say an absence) between the barcode sign, and what it is said to signify, the fish signatory. A space of ambiguity between sign and signified, we argued, needs to be in place so that the ultimate object of concern, not DNA barcodes, but the living breathing, reproducing, dying, renewing and evolving natural biodiversity of living species may become a matter for care, and commitment, as well as the care and commitment for its microgenomic sequencing, naming, classification and documentation.

In this case, everyone involved can see that the stakes and the politics are running high, as DNA-based techniques, following the success of barcoding in certain cases, become a mysterious and compelling black box, capable of conjuring fish in places where they cannot physically be seen. As Darling and Mahon stress, few people understand the workings of these more sensitive detection methods, and those that do are making pleas for a debate that will bring stakeholders up to date on the uncertainties of the sciences being employed (Darling and Mahon 2011). The worst nightmares of barcoding's critics – that it would stimulate and support the development of a cadre of professionals who would need and use barcoding species identification techniques, but who would have very little technical, scientific or taxonomic knowledge – seem to be coming to fruition (Will *et al.* 2005). In addition, the presence of the carp, detected without catching and testing any fish themselves, are capable of igniting a significant political controversy. In the Great Lakes carp controversy, barcoding has reached far out into its identity 'as just a tool' for important environmental monitoring. In doing so it has moved miles away from taxonomy. In these Mode 2 borderlands in which barcoding is developing, science and research are no longer 'terminal or authoritative projects' (Nowotny *et al.* 2001: 2), with determinate outcomes. Rather, 'by creating new knowledge, they add fresh elements of uncertainty and instability' (ibid.). Connections in the pluriverse, to put this in Latour's terms, are extended, but they do not reinstate a Nature that can tell Politics what to do.

Moving into health, leaving taxonomy behind

We now come to our last 'truncated' example, in which the blood contents of the midguts of tsetse flies were analysed using barcoding techniques. It is clear here that this application of barcoding does not add to existing knowledge of the taxonomy of the tsetse flies concerned (the sampled flies were themselves identified to species level through morphological techniques), but it does add to knowledge of their ecology (what different species in different ecosystems feed on), which in turn may inform strategies for tsetse control. The application of barcoding in this instance, therefore, creates and sustains a long chain of connections – from DNA in tiny blood samples, to the prior classification of African mammals, fish and reptiles, to knowledge of ecological feeding chains, and to the future design of public health strategies in the field – all of which could be seen to 'release' the value, or *create new value*, from the past taxonomic classification and identification of African fauna in an important and novel way. The value of this is 'outward' facing: looking towards the world of public health strategies. It does not flow back 'inwards' towards the taxonomic sciences. Barcoding and the BOLD database, in this sense, offer this 'service' to the pluriverse with no 'organizing' in place to support reciprocity. No major plans seem to be in place to barcode all tsetse fly species to aid the slowly building taxonomic classification of these creatures. As the scientist we met on this project conveyed to us in an e-mail: 'Tsetse flies are a difficult taxon, largely because they are spread all over Africa and the species are scattered in different places. It will take some time to complete the exercise'.[10]

The picture we are building up here, through these different applied uses of barcoding, is a picture of shifting relationships within the science of taxonomy, in which the latent value of previous collections, specimens and classification is drawn upon and drawn together, through barcoding infrastructures, and given new life. These barcoding infrastructures are deployed within the pluriversal, Mode 2 borderlands as a black box. As such, they act as a concentration of value, technological know-how, and infrastructure that does not need to be fully understood by the interdisciplinary user-team. The value-laden black box is oriented outwards – towards utility for other social actors, towards releasing value for them, towards opening up circuitry into new outward facing networks. We recall here the sense of sustaining relations across distance that Strathern's correspondent from Papua New Guinea was trying to convey:

> When someone plants a seedling, and that seedling grows into a tree, and the tree bears much fruit, will the eater of the fruit think on the person who planted the seedling?

Strathern reflects:

> 'Thinking on' is an orientation to another person that is meant to summon all the strands of obligation and mutual acknowledgement/recompense that put people into one another's debt.
>
> (Strathern 2004: 61)

The questions appropriate to the pluriversal borderlands of Mode 2 – of responsibility, moral authorship, and the open-ended query 'how are we going to conduct the [barcoding–taxonomy] relationship?' – remain to be asked in our last example. It might be surmised that the crisis of taxonomy into which barcoding pitched itself as a fast, cheap, powerful and ready-to-go saviour in 2003 was such that these questions were not opened up in debate. Conversely, it is clear that value is being created precisely in *letting go* of taxonomy, in releasing barcoding as a method into new experimental domains. Our fourth and last example, told in full below, highlights precisely this point.

'What is out there?' environmental barcoding

Our last example considers the role of barcoding in biomonitoring of a different kind from that of the carp in the Great Lakes. There, the monitoring object, a particular species, was known. In this example the objects of monitoring are not known. We note by way of introduction here that the barcoding being put to use in this example involves far more than the conventional model of the (mere) application of knowledge or methodological tools to new problem settings. Here, the use and application of barcoding necessarily involves new relationships, but it also involves *development* of the science of barcoding and its sense of purpose. Co-production is at work (Jasanoff and Wynne 1998).

We refer to new applications being pioneered from the Canadian Center for the DNA Barcoding of Life at Guelph to develop barcoding for environmental monitoring. This is not necessarily monitoring for the presence of target species (as in looking for the two known species of carp, in the example given earlier), but rather looking into the possibilities that barcoding offers for assessing, comparing, measuring, and even quantifying environmental quality across spaces (different sites) and across time (from one monitoring episode to the next), without identifying species, just analysing diversity using DNA sequencing and barcoding.

Researchers at Guelph are excited about the prospect of articulating the value stored in the infrastructure of BOLD with a need which they anticipate will grow over the next few years – to assess environmental quality along new, species-indifferent lines. As we have seen, the Guelph researchers, designers of BOLD, biologists and informaticians alike, have learned over the past few years to orient their thinking strongly around *future needs* (Chapters 6 and 7) and their intuition is that, in the future, the current value attributed to knowing bio-*diversity* will be accompanied by an even stronger drive to know bio-*quality* and its differentiation over sites and across time. Their sense of this is broadly aligned with larger shifts in biodiversity and conservation policy circles towards ecosystem assessment and 'ecosystem services'.

Given this particular scientific imaginary of the future, many of the questions driving new tools and functions within BOLD itself are clearly related to issues of bio-monitoring: 'What is the species composition of a particular ecosystem?' 'How does biodiversity change over time, space, and in relation to future environmental change?' Hajibabei and colleagues ask (2011). With typically immaculate framing, they note that both questions are difficult to answer in a consistent and timely fashion, and nearly impossible to implement as longer-term monitoring objectives. Species by species analysis for biomonitoring programs and other large-scale biodiversity assessments, especially in multi-species environments like muds, sludges, stream, river, lake or ballast waters, are prohibitive in terms of time and costs, whether a morphological or a DNA-based approach is taken. The density of specimens, running to hundreds to thousands of organisms in a single sample, is simply too high. As an aside here we note how barcoding specifically traded, in its self-promotion, on its comparative advantage of being speedy and cheap, thus offering greater volume of species identification through-put for a given budget or deadline (Hajibabei *et al.* 2005). Yet even so, the mind-boggling numbers involved when real environmental situations are faced, has generated a realization for reasons of bare utility, that taxonomy's defining agenda of species-identification might not always hold sway. Instead, species-indifferent measures of diversity, using genome sequencing and barcoding, begin to look usefull.

Environmental barcoding of unknown bulk specimens using CO1-based (rather than ribosomal based) 454 pyrosequencing is being trialled by Guelph researchers to overcome these problems. CO1-based pyrosequencing is advocated as a way of capitalizing on what Guelph now sees as already established: the known success of CO1 in identifying animal specimens correctly and the robust specimen-to-barcode data accumulating in BOLD's library. On reporting some recent results, we can see that the emphasis is placed not upon species presence, or species diversity per se, but on *comparison* of faunal composition across sites. The methods employed are able to detect habitat variation between sites (e.g. river flow conditions, thermal regime), to infer anthropogenic influence (e.g. chemicals in municipal wastewater effluent, sedimentation from construction

activities in the riparian zone) and to indicate differences in the ecological quality of urban versus conservation habitat (Hajibabei 2011). Identification of every
species is not required.

This kind of research, indifferent to the identification of species or even taxon,
is a short step away from the monitoring of so-called 'traits', where traits are
defined as the 'suite of attributes that a species possesses, with these attributes
influenced by environmental conditions and evolutionary processes' (Culp *et al.*
2011). Traits, in other words, are indicators of environmental conditions, taking
the place of species identification that might have been used in the past to infer
the same. What is currently being proposed is the linking of taxonomic data to the
(collectivized) traits of taxa identified in a particular site. This enables barcoding
to move much deeper into the problem-framing of *environmental biomonitoring*:
differences in the traits of taxa across samples might then be used to generate
additional metrics concerning environmental pressures and their causes upon particular ecosystems (Culp *et al.* 2011).

For both of these possibilities – the use of CO1-based pyrosequencing, and
the linking up of BOLD to traits-based databases – we can see that barcoding is
being extended beyond taxonomy into an imagined future in which biomonitoring
comes to define key elements of the epistemic aims and framing of the environmental sciences. In this imagined future, species definition and boundaries are not
of primary importance: timely and cost-efficient comparison of sites across space
and time, the creation of metrics of quality, and the assessment and quantifica-
tion of these become more salient. DNA-based identification has been shifted to
a more aggregated biological unit than a given species. What we can also see is
that this enterprising and bold anticipation, which also partly involves leading in
the creation of future trends, requires just the same claims of innovation and even
'revolution' (see below) that we initially witnessed in 2003 for barcoding! So far
we have not seen this revolution being extended to encompass a biophilic future
for all of humanity, but a revolution is being promised all the same.

> The ability to automate a biodiversity survey of, for example, bulk macroin-
> vertebrate samples can *revolutionize* large-scale biomonitoring programs that
> are costly, labour-intensive and time-consuming to implement across large
> geographic regions. Moreover, the ability to cheaply and rapidly sequence
> material from different habitats not only increases the efficiency of biomoni-
> toring as a technique, but it expands the scope of monitoring programs, by
> extension into habitats and biota groups which are currently not studied due
> to poor taxonomic knowledge or technical competency.
>
> (Hajibabei *et al.* 2011: 5)

Conclusion

Since barcoding was first presented as a revivalist genomics revolution for taxonomy, and through this for biodiversity and human society, it has driven its
programme forward through much more mundane, localized and practical

achievements in use. In following barcoding into its expanding and diversifying range of uses and user-relations, we reflect a familiar theme in science policy and STS: that while instrumental purposes or promises have always shaped scientific research, in recent decades, demonstrable or plausible future value for society has become a defining normative principle, for its funding and legitimation. Like much of modern genomics, barcoding has 'ridden two horses' with its extravagant visions of future moral transformation and salvation accompanying an emphatic and persuasive programme of useful practical roles for this new genomic tech-noscientific tool. Moreover, in contrast with more prominent genomic projects in the human and biomedical fields (Nightingale and Martin 2004), barcoding has shown that it delivers useful practical results, and this portfolio is expanding.

Barcoding's persuasive emphasis on applications, some of which may have commercializable properties, while also defining itself as a pure/basic science, is one more of the core ambiguities we have traced in this book. But whether barcoding's prime movers see it as able to provide an expanding range of uses to society, with no substantial retroactive influence on the basic science of taxonomy itself, is an open question at present. Here we return to Strathern's Papua New Guinea informant colleague, prompting Strathern to reflect upon the often neglected importance of 'thinking on' – of taking seriously not only particular valuable finite interactions and projects, but crucially, of adopting the moral imagination and responsibility for the open-ended futures of those relationships. The question 'How do we conduct the relationship?' (Strathern 2004: 61) – as a reminder to 'think on' barcoding's continuing debt to the resources and labours of taxonomy's less immediately entrepreneurial scientific field – is certainly being asked and sometimes, if not always, answered. But the conventional long-standing account of the relationship between a notionally pure or basic science, and its offspring, applied science, does not carefully outline the idea of moral debt as Strathern has characterized it. In some cases we can perceive an apparent lack of interest, or capacity, within barcoding, to reflect upon its relations with its key scientific partners, as it cultivates its more immediately compelling relations with users like environmental monitoring agencies, or food regulators. In other cases we see a concern to emphasise barcoding's continuing connections with the sciences and labours of taxonomy (an example here was that of the detection of fish species in restaurants and the reciprocal relationships of this with the Smithsonian's histori-cal and continuing efforts in fish taxonomy).

As our last example in this chapter indicated, more recent developments, stimu-lated almost entirely by its developing relations with external user-worlds, appear to be giving birth to a revolution in barcoding itself, as 'environmental soup' biomonitoring strategies are advertised, tested and developed (Hajibabei *et al.* 2011). This, of course, is a work of co-production. The researchers at Guelph are anticipating users' preferences and needs for the future as they adapt barcoding to new purposes. Their bioinformatics colleagues are simultaneously designing an interface (BINS) that will enable different kinds of metadata to be clustered without needing a definite decision on ultimate identity. BINS is envisaged as a tool that will bring users together to make sense of different data sets before they need to worry about taxonomy. Taxonomy, in effect, may follow on from this, but it does not need to be prior. There is an excitement around barcoding's possible role in environmental biomonitoring in these examples, which are imagined as applications that will have relevance and purchase in the future, bringing non-taxonomist users together around issues of environmental quality at a variety of levels and scales. This excitement also, in part, derives from the liberation of thinking about life, nature and the environment, without the need to 'lock down' species identifications. Nature can be perceived and known in other ways. But since barcoding was meant to be a genomics-based (including informatics) innovation for taxonomy, these extraordinary, and for taxonomy possibly threatening new departures, also raise profound questions about the intellectual, material and social–institutional relationships which sustain and feed barcoding – in the present, but also those that will pertain into the future.

In the final chapter we return to the issue of changing cultures of taxonomy in recent times. One issue which has begun to strike us – especially if we consider Chapter 7's focus on ambitious promises and how might these be validated or held to account, together with this chapter's focus on the rapidly developing horizons of more mundane instrumental uses of barcoding – is that the bases for accountability, and hence legitimation or challenge, are themselves being moved along, in a continually ad-hoc improvisatory manner, with the developing relations of use. The 'revolution' of barcoding's potential role in environmental monitoring now being advertised, which promises to replace the original 'revolutionary' rationale for barcoding as archiving and hence protecting all biodiversity, and taxonomy with it, is a part of this. We 'think on' through this and related issues in the final chapter.

9 Barcoding nature

Final reflections

As we rehearsed in the Introduction to this book, our long-standing colleague at the Natural History Museum in London urged us to embark upon a study of DNA barcoding in the early 2000s because he thought at that time that this innovation looked set to 'rock the foundations of taxonomy'. As a cutting-edge environmental genomics technoscientific innovation combining ambitious promises for the future with what our colleague imagined would be a radical rewriting of taxonomy, BOLI immediately looked worthy of study. But enlarging this focus to consider the daunting global environmental and cultural challenges within which barcoding was emerging – biodiversity loss, taxonomy crisis, 'authority-crisis' for science, and changing practices and scales of knowledge production and use, to put it simply – made his suggestion hard to resist.

Our colleague's initial prediction (that barcoding would revolutionize the taxonomic world) has not actually come to pass in the decade since barcoding was first proposed. DNA barcoding, so very recently hailed as having 'revolutionary' potential, now sits alongside an array of taxonomic methods and plays a modest role within a still fecund ecology of epistemological frameworks for the identification and ordering of life forms. Once we took on board, however, the idea that BOLI aimed from its instantiation to 'make every species count', by using DNA barcoding of species to create a 'bio-literate' global society, we realized that our study would be enriched if we were to explore the societal, ethical and even implicitly spiritual qualities which this genomic technoscientific innovation-in-the-making was promising to bring to late modernity, and to its defining anxieties over nature, knowledge, human meaning and collective purpose. In other words, our research quickly extended well beyond the situated concerns of the sciences of genomics, barcoding and taxonomy, while remaining firmly focused on them.

This escalating extension of focus is a dynamic inherent to the idea of the innovation of barcoding and, like others following genomic actors and their technosciences, we have tried to keep pace with dizzying claims of novelty and purchase, whilst keeping a steady eye for historical continuities and more subtle cultural shifts and processes (Atkinson *et al.* 2009: 12). Since we began drafting this book, innovation, typically identified as solely scientific or technological, has become an 'emergency' project both for economic production and for global salvation.[1] Grand challenges confronting human society in the opening years of the twenty-first century, such as climate change and its human causes; global food security; democratic freedoms and responsibilities; biodiversity loss; synthesizing life itself; human bodily enhancements; and related issues of justice, care, identity, tolerance and equality, have been defined in ambitious but also reductionist terms as challenges which modern technoscience can resolve. Extravagant promises have often appeared in the guise of clear material propositional claims of future beneficial impacts which will be delivered; indeed such claims seem almost a required condition of competitive funding for resource-demanding, technology-intensive research work (Brown *et al.* 2000; Fortun 2009). But of course, when such novel knowledge-programmes meet the complexities and contingencies of real-world experimental enactments, the ostensible clarity and certainty of promises dissolve quietly into flexible diversions, accommodations, improvizations and multiples.

Context is, in this way, absorbed as substantive content to the science, and we have been fortunate enough to witness this happening as it has played out, 'in-action', in the short time that barcoding has been a practised technoscience. At the same time, the apparent clarity of scientific knowledge-production and innovation has been made more complex not only by its diverse material enactments, but also by accompanying, perhaps enabling, economic, political and social forces in the co-production of knowledge and socio-cultural orders (Nightingale and Martin 2004). But as ambition and promises invariably turn into more modest, incomplete and ever-emergent, insecure, and plural practices, *recognition* of such contradictions is often obscured, and the recent history of those now-overambitious promises is forgotten. Through the narrative incrementally developed through the chapters of this book, essential questions implied here about the *public* role of science – what this means and what might need to be done to address its proper enactment – have become more prominent in our exploration of the remit, visions and achievements of BOLI (especially Chapters 7 and 8). We offer additional reflections on this public role further on in this final chapter.

From the short, vibrant, but not uncontroversial history of the Barcoding of Life technoscientific programme, we have described several distinct elements of apparent contradiction arising perhaps inevitably from these entangled and co-dependent conditions of unity, difference and complexity carried by BOLI. The chapters of this book express a number of disconcertments (Verran 2001; 2002) arising from our fieldwork observations of these. As Verran found in her study of Aboriginal and Western scientific land-firing regimes, paying attention to subtly disconcerting encounters in fieldwork seems to open out the potential 'for theorizing, for telling differences and samenesses in new ways' (Verran 2002: 729). In Verran's case, her observations and the disconcertments they generated led to what she calls 'a post-colonial moment' (Verran 2002: 729), a moment where a more abstracted level of analysis became possible, making stark cultural differences suddenly visible, more subtly and deeply comparable, and generalizable in promising new ways. In our case, two broad ambiguities inherent in BOLI which initially provoked in us unease akin to Verran's disconcertments have been converted to curiosity. The way in which what is ostensibly impartial and neutral scientific knowledge can, for example, become normatively embedded as moral imagination and motivator for society has been an important part of our analysis. This issue has troubled science and natural philosophy across a far wider technical and human terrain than only genomics or biology, and over a far longer time-span than just recent decades (Toulmin 1982; Shapin and Schaffer 1985). The particular salvationary programme which BOLI has set in train is just one example of this broader tendency to generate apparent contradiction, and we explore not only the politics but the poetics of this in this final chapter. Such a poetics would, we argue, potentially generate not only a new ethics and politics, but also a new epistemic culture, and a new future, of DNA barcoding, and taxonomy.

We have also described the interesting ways in which the more and more intense ontological entanglements between the human and the non-human, the social and the natural, have necessarily elaborated and thickened as BOLI has embraced and attempted practically to enact its own original ambitions, promises and commitments. Yet, as we have analysed, these particular nature–culture entanglements – such as CO1; a particular bioinformatics algorithm; or the idea of non-commercial science – have often been reformulated to appear, as Latour (1994) originally described, into purely natural universals, self-evident realities, or unconditional standards. Thus the ambiguities that exist in the practical entanglements that are constitutional of the very success of barcoding can be, and are, easily erased. Yet, our point is that we need to keep an eye on these ambiguities, as they have complex lives, and our chapters pursue an analysis and understanding of them as located in specific practices and multiple relationships. Our hunch is that their erasure or forgetting, and the erasure of possible relations and actors which this involves, may entail foregone substantive opportunities and interesting potentials for this technoscience, for its undetermined future forms and relations, and for the publics it envisions and brings into being (Ellis *et al.* 2009).

We come back to these ideas at the end of our reflections. But for now we remind ourselves and our readers of a few of the ambiguities, specific to BOLI, that we have witnessed. The first concerns the phenomenon of standardization. Global standards (for nature and of culture) have understandably been a prominent feature of BOLI, and a conspicuous support for what was initially, at least, hoped to be a single unitary DNA species-marker. Standardization, however, along with its precise intellectual drive, has also dissolved into multiples as barcoding has progressed: the first standard CO1 gene marker for DNA barcoding, for example, had to make way for 'the plant markers', before a flood of further genetic markers for fungi and microbes came to be accepted within barcoding (see the Technical interlude, Chapters 2 and 3). Thus, barcoding's progress – and DNA barcoding itself – was converted into new terms. Such transformations, and the history of barcoding, were unified, however, as if nothing significant had happened. Throughout the book we have noted the ambiguities of what are defined as precise and universal standards, and we have charted the emergence of situational multiplicity, as development has proceeded, and as different human priorities, practices and needs, as well as different natural biological conditions, have been engaged. An interesting point here is that the enlargement of the BOLI programme, which was first promoted as based on CO1 as a universal single standard, arguably gained in robustness as it gained ambiguity and lost its initial singular precision. One might even say that a successful public technoscience like BOLI seems to gain a certain sort of nourishment and enrichment from such contingent accommodation. A question arises for us here, as to whether it also requires the deletion of those contingencies in its substantive representations as public science. A further question also suggests itself: at what point(s) might such a technoscience in fact become unviable, if it were to begin the different historical process of acknowledging those pluralities and contingencies as essential elements of its own practices, and accept the substantively new directional and epistemic influences that this might bring?

The second ambiguity we highlight here concerns the relationship between non-commercial and commercial research. We have noted the ambiguous relationship that BOLI has had, to date, with the potential for commercial exploitation of biological materials and their even less accountable and less governable digitally mobilized derivatives from bio-diverse areas of the world (Chapters 4 and 5). Though BOLI played a role in securing a successful formal resolution on the issue of sourcing biological material for research through the UNCBD Nagoya Protocol, the non-commercial research identity which it needed to adopt to achieve this (as distinct from a more applied/commercial science identity), is also, as BOLI acknowledges, deeply ambiguous. Downstream further possible uses, including commercial uses of such material by other parties, has always been a matter of legitimate concern and controversy, in many parts of the world (Harmon 2010; Reardon and Tallbear 2012; Reardon 2005; Parry 2004). Whatever its own intentions or preferences (and it is important to note here that in our research with BOLI we have seen no evidence of commercial interests which would alarm provider-countries), BOLI cannot claim exemption from future possibilities in this regard, in a world overrun with commercial interest.

A third contradiction concerns the relationship between accumulation and loss (Chapter 6). Here the drive to archive all life, in a single DNA barcoding database, raises cultural questions about human care, control and reduction of the living world and raises the spectre of an impending sense of loss. Just as barcoding technoscience embodies the promise of care, so too has care, and concern about loss, destruction and death featured as a concern of STS and allied scholarship (Haraway 2007, 2010). BOLI seems to be providing the means of redeeming nature and humankind from an abject collective human lack of care to protect life on our planet, whilst paradoxically its actions on this earth seem akin to the impulses of accumulation and control that have led to the same planetary destruction from which it is promising to save humankind.

One way that we have tried to understand these distinct, if overlapping, processes of apparent contradiction has been to focus upon forms of 'forgetting'. The question of whether such acts of amnesia are brought about unintentionally or by active denial we leave as an open one. Thus, for example, we noted how BOLI has been selective in its acknowledgement of its continuing dependencies upon the historical resources of taxonomy whilst energetically reinforcing an imaginary of its own key contributions to the discipline (Chapter 2). We also saw how proliferating public 'user' communities beyond the discipline, ones venturing into a range of applications which not only draw upon taxonomic tradition but also provide ways of transcending its disciplinary commitments and parameters, can either remember their reciprocal ties to taxonomy, or perhaps forget them (Chapter 8). A particular kind of forgetting is also seen in the perpetuation, albeit in a radically different context, of asymmetrical political relations around biodiversity collection, as BOLI continues a colonial tradition of reducing subtle and complex human subjects to rearers and providers of biological material (Chapter 4). BOLI also ostensibly forgets as it wipes the historical biopolitical slate clean through its own self-imaginary as deliverer of humankind's redemption from a destructive future (Chapters 4 and 7). Lastly, forgetting is part of the reality that barcoding's far-reaching promises will never be held to account, since the future to which they refer itself unfolds as newly emergent visions and promises, articulated within a different account of its past.

As we hope will be clear by now, a sense of the apparent contradiction between barcoding's universalizing, revolutionizing ambition and its more mundane achievements, present from the very beginning of our research, has been approached by us not as a particular problem or matter of critique per se, but as something we must learn how to analyse. This analysis of course also influences whatever modest forms of intervention scholars like ourselves might pursue. A recent quip by Mike Fortun, 'there's a crack in everything!' seems relevant here.[2] We note furthermore that it is also not especially difficult to find such contradictions, ambiguities and fissures in grand and hopeful narratives, and we argue here that merely to identify these is *not only* not the point, it is also not the exclusive preserve of ethnographers to do so. As many ethnographers of technoscience have noted, awareness of such contradictions is often demonstrated first by those actually proposing such grand and hopeful schemes: science is not at all incapable of a sense of irony and paradox about its own practices even if subtle taboos are simultaneously in play (Marcus 1995; Fischer 2005; Fortun 2009). We suggest here as a provocation, that perhaps we should not expect that the promissory and other accompanying 'excesses' of a technoscience such as barcoding can ever be exorcised, even as recent political economic conditions have amplified their volume and scope. Perhaps, rather, such excess needs to be seen for what it is, in mutual relation to its also-acknowledged contradiction.

In relation to this point, we see here the importance of a resonance between the STS work of Latour and Haraway, in their acknowledgement of entanglements and their commitment to 'staying with the trouble'. Latour has described the modernist paradox as a dominant drive towards more-and-more complex and deeper entanglements between human and natural entities, a drive which is accompanied by declarations of the purity of the 'natural', and scientific, and of the social, and human, even while these are utterly and irretrievably interwoven. Both authors refer, in different ways, to the existential incoherence between categories (e.g. of nature and culture) which, as part of 'modern' practices, undergo particular forms of purification into the natural and the human. They make the point, in different ways, that that very purification inadvertently forecloses historical potentials for betterment, while the disconcertments, risks and troubles which attend that historical potentiality are thus avoided. No pain, no gain, as-it-were!

To us this tension at the centre of nature–culture entanglements seems to be at least partly connected with the distinction between the private, informal and intimate dimension, and the public, more formal and estranged dimension of technosciences like DNA barcoding and BOLI. In this distinction lies the tension – indeed contradiction – between representing the processes involved in barcoding of natural organisms on the one hand as if purely natural or purely human; and on the other as if they are, as we encountered and observed them, entangled, and in nature–culture classificatory terms, ambiguous. The interesting point is that once we think also of their public, cultural dimensions (as we have begun to do more self-consciously in Chapters 7 and 8), and thus include the issues of authority, credibility, and persuasion of imagined others, thus of larger (public) meanings, then purification becomes not just relevant, but perhaps essential, as a condition of public authority and effect.

Returning to our ethnographic observation and analysis of DNA barcoding and BOLI, it would be unfair however to see our portrayal of these tensions and contradictions as a straightforward critique of barcoding and our BOLI colleagues. They are critical observations about scientific modernity and its cultural formation, but they are not criticisms of actors or their specific actions. Rather, they are our attempt, with the help of STS scholarly insight, and also crucially of the generous collegial openness of BOLI practitioners to our ethnographic interventions, to identify ways by which we might collectively explore a more robust and socially attuned kind of technoscience. In this we resonate with the work of Reardon (2013) and Puig de la Bellacasa (2011) on how a culture of care, and justice, could be given purchase in our own STS work as well as in the technoscientific cultures through whose study we also modestly intervene. Reardon proposes abandonment of the entrenched bioethics paradigm (which has largely failed in its purported role of regulating modern biosciences in a philosophically informed public interest manner), to work with the different one of justice. She essentially builds on the point which STS has worked on but bioethics has neglected, that 'dominant forms of genomics are put together from very particular conceptions of biology, humanity and justice that bring resources and attention to some lives while occluding others' (Reardon 2013). Thus, as we opened out in Chapters 4 and 5, care cannot be detached from material questions of justice; and justice, including ethics, cannot be claimed in the absence of attention to which relations, social groups and non-human entities are valorized, and which are – deliberately or not, regardless – negated and denied recognition.

This includes, as we have indiacted, accepting the new and unfamiliar trouble of sustained attention to the neglected and disempowered human and non-human actors, the obvious and sometimes more subtle dimensions of their subjectivities, their voices and concerns which suffer silencing and deletion in so much prevailing 'responsible' research and innovation, including that which celebrates its own public engagement virtues. Here our case is one of a genomic technoscience already avowedly committed to protection, of nature and of human sustainability; but this same normatively imbued perspective could also be articulated for those predominant genomic others, the more *production-oriented* genomics technosciences. Our general insight here has already been advanced in essence (Haraway 2007, 2010; Reardon 2013; Latour 2004; Fortun 2009), and it is relevant far beyond the already broad reaches of genomics-related domains. It is that, whilst technoscientific development and innovation programmes of all kinds are purified of human or social relations, yet simultaneously utterly entangled in them, they come to be shaped by less deliberate, perhaps less direct, deeper and less obvious but powerful cultural and human imaginaries, and normative influences.

To help us with our final reflections on what we have begun to unfold through the chapters of this book, we turn to a novelist and two anthropologists, each of whom we believe to have original insights concerning how to approach some of the commonplace tensions and contradictions we have highlighted above. In his novel, *The Museum of Innocence*, Orhan Pamuk (2009) describes how a mourner for his lost young lover creates a museum to commemorate his soulmate, and their love, in perpetuity. He promises himself that this archive of lifeless objects, in all their mundane obscurity and triviality, which he had collected from their years of life together, will teach future strangers of a relationship of love, and care, as reflected in his own knowledge of and feeling for every object in the archive. His museum, in Pamuk's words, 'is both a map of a society, and of his heart'. The barcode map of nature which BOLI is busy building, and promising, is of course a very different one. Yet it embodies some similar ambitions of human inspiration, but also ambivalences: of collection and care, even longing, yet with a painful sense of loss, death, high ambition, promise and redemptive meanings projected into and from its own precious objects. But, in Pamuk's case, the archive can only mean something to others if they are also inspired somehow to engender for themselves, the moral-relational imagination to match the mourning lover's affective identifications.

The global collection, production and curacy of BOLD's barcodes of global life, along with its ancillary materials, is pervaded with palpable care and commitment. Yet, as we describe in many of our chapters, the analytical chain from 'wild' nature to its digital database 'map' is long, entangled and fragile, in need of incessant maintenance work. The barcodes in BOLD, as objects of this genomic map of global biodiversity, are unimaginably dense and powerful, semiotically-speaking. We have therefore posed the question: how does the orderly and accountable assemblage of DNA barcodes of life, with all its carefully designed and incorporated forms of accountability, uses, and access, generate a collective sense and practice of care? (Yussof 2011; Puig de la Bellacasa 2011). As the biosemiotics which we discussed using Derrida in Chapters 6 and 7 suggests, the distance, not usually sufficiently acknowledged, between nature and its barcodes is vitally important, as a space of epistemic and ethical tension and indeterminacy in which a moral attitude of care could perhaps be developed, but which is by no means guaranteed. We have already noted that this transformative potential is suggestively articulated in DNA barcoding, as promise; but its chances, and its means of achievement, are vague. Hastrup (1993) has suggested, in such circumstances, that even direct knowledge of loss or suffering, or fragility or threat, let alone distant indirect knowledge, cannot guarantee the condition of care, unless a moral imagination is also somehow articulated (see also Puig de la Bellacasa 2011). This is a challenge for us, as observers *of* barcoders and their work, but we respectfully suggest here that it is also one *for* barcoders themselves, and for the untraced and unmade, as well as the traced, connections they build. Such a moral imagination has to be cultivated. New relationships may be both a consequence, and a cause. If archived objects, for example (here digital barcodes) become tacitly imagined in practice *as if literally* that different and distant natural life in all its unknown glorious variety and generative capacities, then, perhaps, also the genuine culture of care which is invested in the components of the assemblage – human, and non-human – could come to be projected also, onto the wild nature of global biodiversity? This cannot be taken to be a propositional statement about the future, it is just one among other possible imaginaries, but if sincerely taken up it would no doubt change the nature and boundaries of BOLI in radical ways.

We now weave Hastrup's insight concerning the need for a moral imagination together with the suggestively critical observation on the scientific term 'species diversity' itself, by the anthropologist Michael Taussig (2004: 310):

> In my book, phrases such as 'species diversity' belong to a strain of rhetoric better suited to stock-market portfolios than the play of light and water across the rippling rapids of a coastal river. Meant to marshal science in the fight to redeem fallen nature, such phrases actually give a further twist to its destruction. As such, this language takes its place alongside the rhetoric of 'human capital'.[3]

Taussig draws upon the philosophy of Walter Benjamin (1933, 1999) in stressing the importance of always problematizing the relations between words and their things; for example between DNA barcodes and the global nature that they come, or it is promised that they will come, to represent. Introducing the role of the 'poetic' flow of (mis)connections between words, people and things in a way highly relevant for our study, he emphasizes how Benjamin, aware of how powerfully words and pictures 'can engulf and absorb us', nevertheless stressed that 'such connections are intensely playful and political, what he called "allegorical"', by which he meant that the space between a symbol and what it means is always subject to history and is therefore forever incomplete' (Taussig 2004: 312).

Taussig also warned his own trade, anthropologists themselves, not to forget that in making its living by 'telling other peoples' stories', they (we) must not forget to be sensitive to the task of the storytellers themselves – 'We do not have "informants". We live with storytellers, whom too often we have betrayed for the sake of an illusory science' (ibid.: 314). Our own practitioner storytellers were our Barcoding of Life project colleagues, from whose time, patience and willing communications this book has been made possible. We have to interpret their stories, in light of how we understand their multifarious worlds of practice, and with the theoretical resources we can muster. We have identified, *inter alia*, amidst the energetic and confident but flexible process of building BOLI as a global technoscience in service of policy and publics, a preference for celebrating precision and accumulation rather than lamenting the non-achievement of more sublime salvational aims (but this may have been a lack of good questions from us). As part of our final reflection about how such a big technoscientific project could be different, or in other words more likely to succeed in its proposed aim of engaging global publics in a project of caring and knowing, we close by wondering if cultivating a 'poetic' ethos of the intense technoscientific discourse-practices of BOLI might be a key dimension of this, in the way outlined above using Taussig and Benjamin. Might this more poetic culture, if it were established, be one which could embrace its own ambiguities and contradictions as public forms? Might such a poetic ethos also be one in which the normal recognition of others and of difference(s), could be generative of new robust, effective and sustainable public technoscientific culture? Could DNA barcoding embrace such an ethos?

In other essays on Benjamin, Taussig (2006) describes the vain struggles of indigenous storytellers in traditional Latin American peasant agricultures, to make their own poetic embrace of contradiction understood. As he describes, these traditional poets are faced with 'the forward march of machine and chemical-based agribusiness ..., environmental pillage, mass unemployment, breakdown in the courts and police, forced migration to the cities, and the phenomenal rise of violent gangs of young men and women' (ibid.: 60). Under these unpromising conditions, the storytellers are supposed to make sense of the insistent promise of modernization to bring wealth and freedom from deprivation and insecurity. Here, a subtly poetic idiom of ironic scepticism may be a vital resource. But they are not for turning their backs on all that modernity might be, despite the death-dealing brutalities which its existing forms seem to have brought. Their poetic way of relational knowing and being attempts to embrace and articulate multivalency and contingency, and multi-relational ontological being, yet with this an openness, not antagonism, to that instrumentally effective but one-dimensional and depersonalized mode of cosmopolitanism – modern scientific culture (Graeber 2008; Werbner 2008; Stengers 2010, 2011).

Here we can see glimpses of what Taussig is trying to make possible, by recognizing the unimagined thus unexplored potential lying hidden beneath the excesses of modern life's purification and its ontological entanglements (Latour, 2010). So often it is evident that no one believes anyway in those myths of purity – of pure nature, and pure natural knowledge, and of pure culture. Taking the risk of facing head-on the public realities of contradiction, contingency, denial and ambiguity, as well as conditions of faith, and hope, as normal attributes of respectable technoscience, might, we believe, turn out to be less apocalyptic than implicitly feared. Indeed, the lack of control which such a cultural and ethical shift would also embrace, could perhaps be a liberation for technosciences which anxiously seek public authority and effect.

The apparent excesses which we have identified in BOLI's culture: of extravagant promise; of the risk of extreme reification of the DNA barcode as 'nature' in all its generative diversity; of ambiguous claims of pure research innocence; and of effective denial of its own relations of dependency and reciprocity with established partner sciences like taxonomy; can be critically 'exposed' and BOLI's claims thus repudiated. But this is not our conclusion at all. Instead, these apparent denials and excesses could be transformed and read more poetically, as an experimental public articulation of what could perhaps eventually become functional public myths, rather than literally-taken claims of truths of nature or scientific authority. Functionality here – what makes barcoding 'work' – might include: its possible capacity to respect and accommodate difference, and thus a readiness to recognize marginalized and neglected ontologies as legitimate actors in barcoding and global biodiversity; and to acknowledge and embrace, as a public matter, its own contradictions, as part of a continuing, ever-emergent, entangled, indeterminate but organized and effective technoscientific assemblage.

Finally, we suggest that these poetic sensibilities can be woven more closely in to our own disciplinary commitments and interventions. A strong focus in STS and anthropology of science has been to bring to the fore issues of justice, care, and modesty, thus knowledge-ethics, and knowledge-politics, in a world organized to applaud, even to require, epistemic *im*modesty. As the chapters of this book have shown, BOLI actors have, in many instances, embraced immodesty, and DNA barcoding has, not surprisingly, been accompanied by inevitable questions of ethics, justice, politics and power. In suggesting that the immodest excesses of BOLI cannot be treated as ephemeral to it, but have to be understood as constitutive of its effective, and thus far successful life as big genomics-based environmental technoscience, we have read into our case study a kind of dark poetry. But we suggest that this possible poetic sensibility and the openings which this would encourage in technoscience towards multivalent being, and recognition of legitimate difference within its worlds-of-action, can be entwined with the justice-motivated, and modesty-supporting concerns of STS and anthropology of science. In its universalizing quest to be *the* form of taxonomic work that might solve the biodiversity and taxonomic crises combined, barcoding and BOLI can still face questions from sociologists and anthropologists about what more this innovation as history-in-the-making could absorb (and be reciprocally shaped by) along the way. We sociologists and anthropologists might point out the weaknesses, for example, of a largely *presumed* universalism. And we might ask what would happen if barcoding actors thought about DNA barcoding less as a scientizing project, and more as a human-political, even as a collective-experimental ethical project, albeit one crucially informed by state-of-the-art genomics, informatics, and taxonomy. Such questions are not easily answered of course, but they are perhaps a way of encouraging the practice of 'thinking on …' in a spirit of responsibility and relationality, in the manner that Strathern (2004) has articulated (Chapter 8). Such questions may even outline a sensibility which actors and participants from technoscientific, STS, anthropological and other communities alike could each commit to, perhaps with considerable benefits for all. Recognizing, as a thoroughly public matter, ambiguity, contradiction and excess as normal parts of BOLI's story, but also as elements around which things might be done differently, in new relations with new actors, could influence DNA barcoding's future history in remarkable and unforeseen ways.

Notes

1 Introduction

1 This research on biodiversity knowledge networks (2002–2005) occurred before we began to look seriously at the nature of taxonomy itself. It was funded through a three-year research project by the UK Economic and Social Research Council (ESRC). The study was called 'Amateurs as Experts: Harnessing New Knowledge Networks for Biodiversity': http://www.lancs.ac.uk/fss/projects/ieppp/amateurs, accessed 12 March 2013. The research explored differences and convergences in the way that amateur naturalists and professional conservationists in Britain come to know and make records of the natural world.

2 The current research was also funded through a three-year research project by the UK Economic and Social Research Council (ESRC). The study was called 'Taxonomy at a Crossroads: Science, Publics and Policy in Biodiversity'. See http://www.lancs.ac.uk/fass/projects/taxonomy, accessed 12 March 2013.

3 Eukaryotic organisms comprise the plant, animal, fungal and protist kingdoms, but not the two bacterial (eubacterial or archaebacterial) ones, which are prokaryotic. While eukaryote cells have a nucleus, and have mitochondrial DNA exogenous to their nuclear DNA, prokaryotes do not have a nucleus, nor do they have mitochondria. As explained in the Technical interlude, the standard gene fragment originally selected for barcoding, from the CO1 gene, is only found in mitochondrial DNA.

4 The Linnaean hierarchy is the organization of life into the successively subordinated eight taxonomic ranks – Domain; Kingdom; Phylum; Class; Order; Family; Genus; Species.

Technical interlude

1 Taken from IBOL, CBOL and BOLI websites: http://ibol.org; http://www.barcodeoflife.org/content/about/what-cbol; http://www.barcodeoflife.org, accessed 23 March 2012.

2 DNA barcoding: revolution or conciliation?

1 See Franklin (2005: 74) for a similar attempt to think about the 'culture' in which post-genomic research is fed and nourished.

2 Although studies 'testing' the applicability of CO1 analysis on different taxa featured from the early 2000s and continue to populate the literature on a case study basis.

3 This delicate balancing act is also described in other STS case studies, see, for example, Clarke and Fujimura (1992).

4 Latour, B. (1987) *Science in Action: How to Follow Scientists and Engineers Through Society*. Cambridge, MA: Harvard University Press.

5 Lucy Suchman writes about the 'premium placed on discrete, discontinuous change events in projects of innovation' (Suchman and Bishop 2000: 332; see also Suchman 1994, 2007).

6 Hebert is reported as believing that 'There is no more likely death of a discipline than the failure to innovate' (Paul Hebert, quoted by Nicholls (2003)). Please re-word footnote to say: Nicholls, H. 2003. DNA: The barcode of life? Originally published on behalf of Elsevier by BioMedNet. Now available at http://www.uoguelph.ca/~phebert/media/BioMedNet%20News%20article.pdf, accessed 12 March 2013.

7 Interview with Paul Hebert, Canadian Center for DNA Barcoding, Biodiversity Institute of Ontario, University of Guelph, 2006.

8 See http://www.ccdb.ca/pa/ge/dna-barcoding/barcode-of-life-initiative/barcoding-basics, accessed 15 June 2012.

9 Interview with Researcher, Canadian Center for DNA Barcoding, Biodiversity Institute of Ontario, University of Guelph, November 2006.

10 See Vogler and Monahan (2006) for an excellent depiction of this method/philosophy in contrast to DNA taxonomy.

11 This difference is usually described as the difference between 'intra-species' and 'inter-species' variation.

12 Interview with Researcher, Canadian Center for DNA Barcoding, Biodiversity Institute of Ontario, University of Guelph, 2006.

13 The Banbury meetings as they came to be called had the formal titles of: 'DNA and Taxonomy', Cold Spring Harbor Banbury Conference Center, New York, NY. 9–12 March 2003. 'Taxonomy, DNA, and the Barcode of Life'. Cold Spring Harbor Banbury Conference Center, New York, NY, 10–12 September 2003. Both were sponsored by the Sloan Foundation and were summarized in the following report: http://www.barcodeoflife.org/sites/default/files/legacy/pdf/Banbury2Report.pdf.

14 As Godfray and Knapp 2004 reported, in 2003 the NSF and the All Species Foundation announced the funding of four projects under the Planetary Biodiversity Inventories Programme. The results of these projects were to be web-available taxonomies of four major groups on a global scale. These were not going to be just lists of names, however, they were intended to be content rich and include descriptions and links to other sites (Godfray and Knapp 2004: 564).

15 Interview with Paul Hebert, Canadian Center for DNA Barcoding, Biodiversity Institute of Ontario, University of Guelph, 2006: p. 21.

3 What's in a barcode?: the use, selection and (de)naturalization of genetic markers

1 Interview with Paul Hebert, Canadian Center for DNA Barcoding, Biodiversity Institute of Ontario, University of Guelph, November 2006.

2 Interview with Plant Systematist, Natural History Museum, London, October 2006.

3 Interview with Researcher, Canadian Center for DNA Barcoding, Biodiversity Institute of Ontario, University of Guelph, November 2006.

4 The imagined enthusiasm of a global public re-connected to nature through barcoding is the subject of an earlier analysis in a paper by the authors, Ellis *et al.* (2010) and is also the subject of later analysis in this book in Chapter 7.

5 We use the term 'fabrication' of CO1 deliberately here to highlight the artefactual nature of this segment of the gene, a tiny fraction of the genome, seen as the key to the identification of all animal life.

6 This quest to see barcoding through as a 'global standard in taxonomy' is the main goal of the Consortium for the Barcoding of Life (CBOL), as we shall describe in Chapter 5.

7 We focus here in this chapter on the introduction of 'plants' and 'botanists' to barcoding's story. This is a simplification on our part. We could also have added reflection on the work of taxonomists working on other organisms that were not amenable to species identification through CO1 – such as fungi and protists.

8 There is detailed documentation of the negotiation of the search for markers for fungi and protists. One of the clearest accounts of decisions taken can be found on the website of the Canadian Centre for DNA Barcoding (CCDB): http://www.ccdb.ca/pa/ge/research/domains-of-life, accessed 3 July 2012.

9 GenBank® is a genetic sequence database, run through the USA's National Center for Biotechnology Information (NCBI). It consists of an annotated collection of all publicly available DNA sequences, including those in BOLD.

10 The attributes of *cytochrome oxidase 1* are very clearly discussed on the Canadian Centre for Barcoding Website: http://www.dnabarcoding.ca/primer/Index.html, accessed 12 March 2013.

11 Interview with Paul Hebert, Canadian Center for DNA Barcoding, Biodiversity Institute of Ontario, University of Guelph, November 2006.

12 The 'phenetic' approach is an attempt to classify organisms based on overall similarity, usually in morphology or other observable traits, regardless of their phylogeny or evolutionary relations (Hull 1998).

13 Interview with Paul Hebert, Canadian Center for DNA Barcoding, Biodiversity Institute of Ontario, University of Guelph, November 2006.

14 A primer is a strand of nucleic acid that serves as a starting point for DNA replication.

15 Rationale, methodology and results of the first feasibility study are presented in Hebert *et al.* (2003a). Our account of the study, represented here, combines a reading of the published paper together with information provided on the Canadian Centre for Barcoding Website: http://www.dnabarcoding.ca.

16 See Hebert *et al.* (2003b) for a fuller discussion of the thresholds chosen to decide the acceptability of sequence divergence.

17 See the Technical interlude for a list of basic scientific reasons underpinning the choice of CO1 for global barcoding.

18 Hebert and co-authors' 2003 publications were under review from 2002 and both published online in 2003.

19 The co-shaping of research questions and the agendas of funders has a significant bearing on the trajectory of science, technologies and innovation as Wynne 2007 explores.

20 Interview, Paul Hebert, Canadian Center for DNA Barcoding, Biodiversity Institute of Ontario, University of Guelph, July 2007.

21 Interview, Robyn Cowan and Mark Chase, Royal Botanical Gardens, Kew, July 2007.

22 For full and nuanced accounts of the pros-and-cons of different candidate genetic regions for plants, see a range of publications, including Kress *et al.* (2005); Kress and Erickson (2007, 2008); Little and Stevenson (2007); Pennisi (2007); Hollingsworth *et al.* (2008); Hollingsworth (2008); Lahaye van der Bank, *et al.* (2008); Ledford (2008); Newmaster *et al.* (2008).

23 The plastid genome is commonly called the 'chloroplast' genome.

24 Interview Robyn Cowan and Mark Chase, Royal Botanical Gardens, Kew, July 2007.

25 See Plant Working Group Website: http://www.barcoding.si.edu/plant_working_group.html, accessed 12 March 2013.

26 The aura of competition we highlight here in suggesting the race-like quality of the search for the plant barcode is not entirely dissimilar (although quite different in scale and implication) to the competitive drive to complete the sequencing of the entire human genome. See Reardon (2005).

27 We are grateful to Peter Hollingsworth, Royal Botanical Gardens, Edinburgh, for providing us with this illustration, and permission to reproduce it. He used it as part of a presentation to the barcoding community in San Diego, July 2008.

28 Interviews with Robyn Cowan and Mark Chase, Royal Botanical Gardens, Kew, July 2007 and Peter Hollingsworth, Edinburgh Botanical Gardens, July 2007.
29 DAWG is one of the working groups established as part of CBOL: http://www. barcoding.si.edu/working_groups.html, accessed 12 March 2013. According to CBOL's website; the mission of this new working group is to explore and develop ways of analysing barcode data for the purpose of assigning individuals to species, discovering new species, and other appropriate purposes.
30 The region first proposed by Kress *et al.* in 2005 and advocated by Kress in Taipei, for example, was *trnH-psbA* which is a 'spacer' sequence between two genes. This region does not code for protein (synthesis) and it is, as argued by Kress' critics, notoriously unwieldy and difficult to extract, amplify and align. This region has subsequently been the centre of considerable attention and effort on the part of DAWG.
31 Field notes, DAWG meeting, San Diego, June 2008.
32 Dr Kim, from the Research Institute of Basic Sciences, Pusan National University, South Korea, had not hitherto been part of the debate (see Figure 3.1).
33 This resonates with work exploring the 'mereographic' qualities of genomic science, see Franklin 2003; Thompson 2005; Waterton 2010.

4 'A leg away for DNA': mobilizing, compiling and purifying material for DNA barcoding

1 BOLD website http://www.boldsystems.org, accessed 25 August 2012. It is worth noting that although 1,733,466 barcode sequences are currently archived, BOLD states that it currently houses 1,971,659 sequences, indicating an 'excess' of over 200,000 species barcodes which do not yet adhere to BOLI standards.
2 This formally culminated at the Conference of the Parties, in Nagoya 2010, with The Nagoya Protocol on Access to Genetic Resources and the Fair and Equitable Sharing of Benefits Arising from their Utilization of the UN CBD. The Protocol is a transparent framework of international law (for its signatories) established to regulate the fair and equitable sharing of benefits arising out of the use of genetic resources.
3 This is a transformation Parry (2004) explores in her account of the commercial 'microsourcing' of genetic material by pharmaceutical companies of taxonomic collections curated in herbaria and museums.
4 Holloway (2006: 1).
5 The matter of the 'final' taxonomy of *Astraptes fulgerator* is a rather controversial one. Following the publication of Janzen and Hebert's findings (Hebert *et al.*, 2004), Andrew Brower formally described the 10 species without the consent of the original authors (Brower 2010). The Barcoding of Life Database (BOLD) has not adopted the names Brower assigned the interim taxa as the original authors have not supported the move.
6 Hayden (2003) and Parry (2004) refer to parataxonomists in the context of the Merck-INBio flagship project in Costa Rica, and its pioneering bio-prospecting arrangement; but their reference is not part of their broader analytical critique of the practices of collecting and mobilizing biological materials. See Zebich-Knos (1997) for an account of the Merck-INBio agreement as a quintessential 'neoliberal' framing of sustainable development and rainforest conservation.
7 Janzen's publications acknowledge the contribution of 'parataxonomists' to his research as part of a recognition of their broader participation as part of the workforce of INBio (Instituto Nacional de Biodiversidad, Costa Rica: http://www.inbio.ac.cr/en, accessed 12 March 2013).
8 See Fairhead and Leach (1996) for further examples of the tendency to allocate blame in this way.

9   See http://www.barcodingbirds.org, accessed 30 May 2012.
10   In the past few decades, the Smithsonian has played a central role in the growing effort to systematically gather data on bird strikes, this data is then used by the US Air Force and the U.S. Federal Aviation Administration (FAA) to analyse bird behaviour, dietary preferences and migratory patterns using the information to reduce hazards at airports.
NMNH Natural History Highlight: CSI for Birds – Scientists Use Forensic Techniques to Improve Airport Safety: http://www.mnh.si.edu/highlight/feathers, accessed 30 May 2012.
11 Barcoding Life Takes Flight: All Birds Barcoding Initiative (ABBI) Report of Inaugural Workshop. Museum of Comparative Zoology, Harvard University. Cambridge, MA, 7–9 September 2005.
12 Interview with Curator of the Bird Collection of the American Museum of Natural History, New York, 5 February 2008.
13 Interview with data manager, Laboratory of Analytical Biology, Smithsonian Institution, 16 July 2007.
14 The reviewers of this publication were keen to remove phylogenetic information which was in fact initially included. Control was asserted by the reviewers and this was a process the museum curators were aware of and very sensitive to. Interview with Curator of the Bird Collection of the American Museum of Natural History, New York, 5 February 2008.
15 Now the Nagoya Protocol of the CBD: http://www.cbd.int/abs/about, accessed 12 March 2013.
16 We note that the terms 'pure', 'basic' and 'non-commercial' research tend to be used interchangeably in the CBD/ABS/Nagoya Protocol literature. Our own writing further reflects this interchangeablility although we prefer to use the term 'non-commercial' research where possible.
17 The website for the Bonn (2008) COP Access and Benefit Sharing workshop, including the report and subsequent documents, is at http://barcoding.si.edu/ABSworkshop.html, accessed 12 March 2013.
18 CBOL's position on non-commercial research within the CBD was further clarified and potentially consolidated with the timely publication of a Nature Commentary just before the Nagoya COP (see Schindel 2010). The article pleaded for the distinction between non-commercial and other research using materials from biodiversity-rich global countries, to be reflected in the recently added draft text which would free non-commercial research such as BOLI's from regulatory requirements which could delay access thus research by several years, if not indefinitely. The article also explicitly recognized the potential for what initiates as apparently non-commercial research to develop future commercial potential, a possibility which it argues should also be controlled.
19 Reflecting this enlarged global policy interest stimulated by the CBD–ABS entanglements, and its own global growth, iBOL also established a research team to focus on 1) identification of types and origins and materials and associated knowledge to be used in iBOL research; 2) comparative legal review of international and national treaties; 3) interviews with iBOL researchers on ABS issues so as to examine iBOL's policies and practices on data, intellectual property, and material transfers, and to compare these with those of other stakeholders (e.g. industry, NGOs, IPS); 4) an evaluation of indigenous communities' 'attachment' to materials and associated knowledge, iBOL website: http://ibol.org/ibol-cbd, accessed 4 June 2012.
20 For a slightly ominous but realistically cautious legal analysis/assessment of the position of non-commercial research, in which BOLI accompanied many taxonomists and other scientists with long experience of such uncertain and slow access negotiations, see Kamau et al. (2010).

21 The relevant piece of agreed text reads as follows: Article 8: Special Considerations: In the development and implementation of its access and benefit-sharing legislation or regulatory requirements, each party shall:

> Create conditions to promote and encourage research which contributes to the conservation and sustainable use of biological diversity, particularly in developing countries, including through simplified measures on access for non-commercial research purposes, taking into account the need to address a change of intent for such research.

22 iBOL website: http://ibol.org/ibol-cbd-pledge-to-work-together-towards-common-goals, accessed 4 June 2012.

23 Subsequently, iBOL has ostensibly developed a greater awareness of, and sensitivity to the historically rich and troubled relationship between global indigenous peoples and the CBD by defining one of its research foci to 'evaluate indigenous communities 'attachment' to materials and associated knowledge' (see note 19 above). In this context, it should not be forgotten however that most often a definition of *useful knowledge* in relation to biodiversity conservation is limited to that of direct perceived relevance to biodiversity science (e.g. species identification and location). In other words, the local knowledge referred to by iBOL's research focus, even if recognized, is likely to be reduced to scientized data (Agrawal 2002).

24 A position underscored by the UN Declaration on the Rights of Indigenous People (see previous note).

25 See www.ipbes.net, accessed 12 March 2013.

5 Extending the barcoding frontier

1 CBOL has a Scientific Executive group and the members of this group are those that conduct BOLI/CBOL meetings. In this chapter we use the term CBOL executive member for a number of individuals within that group. We do not believe it is important to the understanding of the text to identify individuals within the CBOL executive members and we use this term to protect anonymity.

2 As part of our ethnography of the emergence of barcoding, from 2005–2009, we have followed the activities of CBOL, and been able to attend many meetings and workshops it has organized across the globe, from the very first meeting in the Natural History Museum in London in 2005 to those taking place in Nairobi (Kenya), Washington (USA), Campinas (Brazil), San Diego (USA), Taipei (Taiwan) from 2006–2009.

3 One aspect of barcoding's connection with the co-creation of life forms here refers to the links being made between barcoding and GM crops (Nzeduru 2012). Other links are being made with GM mosquitoes (Beisel 2010).

4 CBOL executive speaking to delegates and the CBOL Leading Laboratories Workshop, San Diego, 23 June 2008. Notes taken at above workshop. As the Nagoya Protocol on ABS Agreement was established more recently, in 2011, this commitment has now been taken over by more recent agreements. However, the example still shows the way that CBOL approached the issue of access to biological material and the way that it interpreted 'benefits'.

5 When we finished this manuscript in 2012, iBOL was indeed suffering from reduced funding input.

6 See http://www.barcodeoflife.org/content/about/barcoding-landscape, accessed 21 November 2011.

7 Notes from our observation of CBOL's South America Regional Meeting, 19 March 2007.

8 Notes from our observation of CBOL South America Regional Meeting, 19 March 2007. A similar view was repeatedly expressed at CBOL international meetings, most notably in South America and South East Asia.

9 Notes from our observation of CBOL South America Regional Meeting, 19 March 2007.
10 Notes from our observation of CBOL South America Regional Meeting, 19 March 2007.
11 In this respect the example given here of South American taxonomists resisting the assumptions of US biologists differed markedly from that given by Sunder Rajan (2006) who notes a deference to US norms and patterns of 'labour' amongst Indian volunteers in pharmaceutical medicine trials.

6 Archiving diversity: BOLD

1 This is no more than an extension of the finitist understanding of science explained, for example by Barnes *et al.* (1996).
2 Environment Canada is a government department responsible for the protection of the environment across Canada. Genome Canada (GC) is a formally independent not-for-profit agency established by the Canadian government in 2000 to promote genomics research and applications.
3 Interview with Paul Hebert, Canadian Center for DNA Barcoding, University of Guelph, November 2006.
4 Interview with Paul Hebert, Canadian Center for DNA Barcoding, University of Guelph, November 2006.
5 High dimensional genetic data are those typically generated by technologies such as microarrays. Though high-productivity, they tend to be 'noisy' data, with potentially thousands of variables, requiring much interpretation by statisticians/bioinformaticians.
6 These national-level databases drew in turn upon earlier smaller laboratory-based practices for curating DNA sequences, for example that of Margaret O. Dayhoff (1925–1983) at the National Biomedical Research Foundation (NBRF) as well as that of a group of researchers around Walter Goad (1925–2000) at Los Alamos National Laboratory, see Strasser (2008); Cook-Deegan (1994); Strasser (2011).
7 A large literature on the emergence of biology as an 'information science' and as a digital science is reviewed in Chow-White and Garcia-Sancho (2011).
8 We note here that the formation and filling of such digital databases with the right information involves much carefully designed research in itself. In biomedical fields, however, these databases have been seen as *platforms* for further research yet to be identified and pursued. As a result the extensive and continuing research on the input-architecture side, which is where BOLI mainly works, is easily deleted from recognition. This may reflect one aspect of the so far underexplored differences for biodigital technoscience cultures, between medically oriented, and environmentally oriented, R&D and innovation.
9 Kwa (2009). In this article Kwa notes the shift of scientists like Venter from being a traditional 'experimentalist' working on the genome of one organism to working in a more 'taxonomic' style – mapping variation, collecting variation (e.g. of a genome, or of the genome of a specific habitat (the deep ocean) and setting it out in a grid, in order to ascertain pattern and relationship.
10 Discussion between Brian Wynne and two Mexican plant scientists, UNAM, Mexico City, 23 May 2011.
11 The idea of 'care for the data' in genomics is a focus of ongoing research and a book project on genomic approaches to understanding asthma by Mike Fortun, http://www.fortuns.org/?page_id=72, accessed 12 March 2013. We have also benefited from several conversations with Ruth McNally on this topic of curacy of data, from her Cesagen work on proteomics.
12 Mackenzie notes that databases are often in the business of creating and curating or mediating multiples:

> Multiples are increasingly mediated through databases and data-base driven architectures. Whilst multiples in the world remain innumerable, databases have increasingly emerged as a way of collecting, enumerating, and enunciating multiples in particular ways. The existence of vast databases (such as Google or Amazon or GenBank) and the existence of huge numbers of databases (including many personal databases that would not be visible or obvious as such to the people who rely on them to manage their gadget equipped lives), I would argue, can be read as a ramification of particular historical renderings of multiples. Databases are a situation in which we materially encounter the doing of the multiple.
>
> (Mackenzie 2011: 4)

13 Interview with researcher at the Canadian Center for DNA Barcoding of Life, University of Guelph, November 2006.

14 This was the the Ebbe Nielsen Prize, awarded in 2010. BOLD was judged by GBIF's science committee to be a 'major and innovative landmark in bringing genomic data on biodiversity to research and research applications for science and society'. Report available at http://www.ibol.org/bold-leader-wins-ebbe-nielsen-prize, accessed 28 March 2012.

15 Interview with researcher at the Canadian Center for DNA Barcoding of Life, University of Guelph, November 2006.

16 See http://ibol.org/resources/barcode-library, accessed 24 April 2012.

17 Interview with researcher at the Canadian Center for DNA Barcoding of Life, University of Guelph, November 2006.

18 Interview with the principal designer of BOLD, Canadian Center for the DNA Barcoding of Life, November 2006.

19 See Lynch, Cole and McNally (2008) on this idea of reversibility of accounting in DNA forensics. Latour's well-known paper 'The Pedofil of Boa Vista' (1999) is one of the clearest articulations in STS of this notion of 'reversibility' of evidence and translation through supply-chains in the sciences extending from field to lab, and back. See Lynch and McNally (2005), Lynch *et al.* (2008) and Waterton (2002) for accounts that complicate this notion of 'going back' reversibly through these nodes, to a static data-point in order to trace connections back to their (supposedly stable and lasting) origins.

20 A 'BARCODE' record (always written in capitals) is one that has met the standard requirements for data entry.

21 The presence of a STOP codon in the middle of a protein encoding region is said instantly to highlight problems in base calling (Humphries *et al.* 2005: 13).

22 These Data Analysis Working Group (DAWG) meetings were held many times over the course of our field work at different locations depending on where taxonomists and bioinformaticians could be brought together to discuss common barcoding issues and problems. The exchange we describe was observed in Taipei, September 2007, both in a plenary session and in a following session devoted to bioinformatic solutions for forthcoming BOLD data issues.

23 These are the Neighbour-joining trees, referred to earlier in Chapter 6 and in the Technical interlude.

24 Mackenzie (2011) has written about the way that algorithms are subsumed in tectonic plate like processes and how strange it is that major shifts in the ontology of the world can be swallowed under in a way that is seemingly imperceptible to many of us living in that world.

25 Comments on this issue e-mailed to the authors, from the principal designer of BOLD, Sujeevan Ratnasingham, CCDB, University of Guelph.

26 This 'freedom' is of course qualified by BOLD's designers' need to intuit also for the future, how any future design is going to give BOLD the capacity to work for the world of specialist, wider policy and public users. This 'working', as McKenzie (1989) has noted,

is itself a social product, as well as a (flexible) function of a technology's correspondence with nature. Our point is that BOLD designers do define what it means for BOLI to work, by tacitly influencing what users define as their needs.

7 BOLI as redemptive technoscientific innovation

1 Paul Hebert, invertebrate taxonomist, Director of the Canadian Center for DNA Barcoding, Biodiversity Institute of Ontario, and Director of iBOL, letter to research team, 4 November 2004.
2 Interview with Paul Hebert, Canadian Center for DNA Barcoding, Biodiversity Institute of Ontario, University of Guelph, Canada, November 2006.
3 Kellert, Stephen R. and Wilson, Edward O. (1993) *The Biophilia Hypothesis.* Washington, DC: Island Press.Wilson, E.O. (1984) *Biophilia: The Human Bond with Other Species.* Cambridge, MA: Harvard University Press. Wilson, E.O (1991) 'Arousing Biophilia: A Conversation with E. O. Wilson', *Orion* 10 (Winter): 9–1.
4 We are aware here of a broad literature on Biosemiotics (e.g. Barbieri 2008; Hoffemeyer 1996). This is a literature which develops from the premise that all living beings, human and non-human operate within systems of signification. What we do not do here and hence are not orthodox 'followers' of biosemiotics is delve into the signing properties of organisms per se – rather we expand the context to observe BOLI as a technoscientific bio-semiotic system which enrolls organisms, technology and human communities in processes of signification, understanding and interpretation.
5 Although we have found ourselves alternating here between the terms 'signature' and 'sign', we have eventually opted, for consistency's sake, to use the term 'signature' (see Helmreich 2009 for a similar use of Derrida's understanding of 'signature' in his analysis of 'bio-signatures').
6 By suggesting that biodiversity is a signatory, we could be seen to be attributing biological organisms with communicative intention. We could think about this in theoretical terms here and enjoy thinking about the agency of species in 'signing' for their own lives. But we do not take up this challenge here. All we suggest is that the DNA barcode as a socio-technical artifact works to connect humans and biodiversity by operating as a readable signature. To all intents and purposes, the organisms themselves are passive in this process and their 'reading' as part of the 'book of nature' is facilitated by the barcode.
7 See Derrida (1988, 1997) and Kirby (2011).
8 Interview with Researcher, Canadian Center for DNA Barcoding, Biodiversity Institute of Ontario, University of Guelph, November 2006, emphasis added.
9 See Derrida's use of Condillac's insistence upon continuity between signature and signatory as a foil against which to explain the importance of a distance and indeterminacy shaping the very technologies of communicative systems (1988).

8 What is it?: identifying nature and valuing utility

1 Nature is given a capital N here to denote the idea that it can stand in, as an uncontested matter of 'fact', and as such can direct formal 'Politics', which is given a capital P to denote its own formal institutional basis. This formulation is laid out most clearly in Latour (2004).
2 Barcoding's main proponents, we recollect, have been located in publically funded institutions – the Smithsonian Museum and the University of Guelph – institutions that have learnt fast to re-orient to this context. 'Bringing in' exterior funding sources to support useful, 'public good' research has become second nature (i.e. a survival strategy) to these institutions within a relatively short time span.

3 We note here that the very idea of public good requires an understanding of historic and cultural context. In a world increasingly influenced by dominant economic-liberal culture, private commercial benefit is equal to public good. Wealth-creation, however you do it, is public good.

4 See http://www.popsci.com/science/article/2010-10/scientists-deploy-dna-forensics-protect-overhunted-animals, accessed 12 March 2013.

5 See http://www.popsci.com/science/article/2010-10/scientists-deploy-dna-forensics-protect-overhunted-animals, accessed 12 March 2013.

6 Interview by the *Boston Globe* with Lee Weight of the Smithsonian Museum: http://smithsonianscience.org/2011/12/smithsonians-work-with-dna-barcoding-is-making-seafood-substitution-and-mislabeling-easier-to-catch, posted 21 December 2011, accessed 12 March 2013.

7 FISH-BOL is co-chaired by Australian Bob Ward and Canadian Paul Hebert with a direct 'portal' to BOLD at Guelph: http://www.fishbol.org/vision.php, accessed 12 March 2013.

8 The following description of the background and the case itself is taken from Darling and Mahon 2011: pages 980–981, and modified only lightly by the present authors, as Darling and Mahon's telling of the story of Asian carp could not be improved upon.

9 Interview by the *Boston Globe* with Lee Weight of the Smithsonian Museum: http://smithsonianscience.org/2011/12/smithsonians-work-with-dna-barcoding-is-making-seafood-substitution-and-mislabeling-easier-to-catch, posted 21 December 2011.

10 E-mail sent to Claire Waterton, 15 May 2012.

9 Barcoding nature: final reflections

1 Maire Geoghegan-Quinn Commissioner for Research, Innovation and Science Innovation Union Scoreboard: opening remarks on the innovation emergency in Europe at the press conference on the European Commission *Horizon 2020* R&D Programme. Brussels, 1 February 2011. See http://europa.eu/rapid/pressReleasesAction.do?reference=SPEECH/11/60&type=HTML , accessed 11 July 2012.

2 Seminar with Mike Fortun speaking on 'Care, Creation and the Impossible Sciences of GeneXEnvironment Interactions in Asthma: Promising Genomics' v.2. Cesagen Seminar, Lancaster University, 9 November 2011. See also Fortun (2009).

3 He might have added here, of 'ecosystems services', as monetarized for free market exchange, and a now dominant means of knowing 'biodiversity'.

Bibliography

Agrawal, A. (2002) 'Indigenous knowledge and the politics of classification', *International Social Science Journal*, 54 (173): 277–281.

Allander, T., Emerson, S.U., Engle, R.E., Purcell, R.H. and Bukh, J. (2001) 'A virus discovery method incorporating DNase treatment and its application to the identification of two bovine parvovirus species', *Proceedings of the National Academy of Sciences of the United States of America*, 98: 11609–11614.

Appadurai, A. (1996) *Modernity at Large*, Minneapolis: University of Minnesota Press.

Arnot D.E., Roper C. and Bayoumi R.A. (1993) 'Digital codes from hypervariable tandemly repeated DNA sequences in the Plasmodium falciparum circumsporozoite gene can genetically barcode isolates', *Molecular and Biochemical Parasitology*, 61: 15–24.

Atkinson, P., Glasner, P. and Lock, M. (2009) 'Genetics and society: perspectives from the twenty-first century', in Atkinson, P., Glasner, P. and Lock, M. *Handbook of Genetics and Society: Mapping the New Genomic Era*, London: Routledge, pp.1–14.

Atran, S. (1985) 'Pre-theoretical aspects of Aristotelian definition and classification of animals', *Studies in History and Philosophy of Science*, 16: 111 162.

Avise, J. (1994) *Molecular Markers, Natural History, and Evolution*, New York: Chapman & Hall.

Avise, J. (2004) *Molecular Markers, Natural History and Evolution*, 2nd Edition, Sunderland, MA: Sinauer Associates.

Ballard, J. and Whitlock, M. (2004) 'The incomplete history of mitochondria', *Molecular Ecology*, 13 (4): 729–744.

Barad, K. (2003) 'Posthumanist performativity: toward an understanding of how matter comes to matter', *Signs: Journal of Women in Culture and Society*, 28(3): 801–831.

Barad, K.M. (2007) *Meeting the Universe Halfway: Quantum Physics and the Entanglement of Matter and Meaning*, Durham, NC: Duke University Press.

Barbieri, M. (2008) 'Biosemiotics: a new understanding of life', *Naturwissenschaften*, 95: 577–599.

Barnes, B. (2010) 'Afterword', in Parry, S. and Dupre, J. *Nature After the Genome*, Oxford: Blackwell.

Barnes, B. and Shapin, S. (eds) (1979) *Natural Order: Historical Studies of Scientific Culture*, London, California, New York: Sage Publications.

Barnes, B., Bloor, D. and Henry, J. (1996) *Scientific Knowledge: A Sociological Analysis*, Chicago: University of Chicago Press.

Barry, A. (2001) *Political Machines: Governing a Technological Society*, London and New York: Athlone Press.

Basset, Y., Novotny, V., Miller, S., Weiblen, G., Missa, O. and Stewart, A. (2004) 'Conservation and biological monitoring of tropical forests: the role of parataxonomists', *Journal of Applied Ecology*, 41: 164–174.

Bataille, G. (1991a) *Accursed Share, Vol. 1*, trans. R. Hurley, New York: Zone Books.

Bataille, G. (1991b) *Accursed Share, Vols 2 and 3*, trans. R. Hurley, New York: Zone Books.

Beck, U. (1992) *Risk Society: Towards a New Modernity*, London, California, New Delhi: Sage Publications.

Beck, U. (1999) *World Risk Society*, Cambridge: Polity Press.

Beisel, U. (2010) 'Jumping hurdles with mosquitoes?' *Environment and Planning D: Society and Space*, 28 (1): 46–49.

Benjamin, W. (1933) 'On the mimetic faculty', in Demetz, P. (ed.) *Reflections: Essays, Aphorisms, Autobiographical Writings*, trans. E. Jephcott, New York: Shocken Books, pp.333–336.

Benjamin, W. (1999) 'The collector', in Tiedemann, R. (ed.) *The Arcades Project 1982*, trans. H. Eiland and K. McLaughlin, Cambridge, MA: Harvard University Press, pp.203–211.

Bensasson, D., Zhang, D-X., Hartl, D. and Hewitt, G. (2001) 'Mitochondrial pseudogenes: evolution's misplaced witnesses', *Trends in Ecology and Evolution*, 16 (6): 314–321.

Bisby, F.A., Shimura, J., Ruggiero, M., Edwards, J. and Haeuser, C. (2002) 'Taxonomy, at the click of a mouse', *Nature*, 418: 367.

Blaxter, M. (2003) 'Counting angels with DNA', *Nature*, 421: 122–124.

Blaxter, M. (2004) 'The promise of a DNA taxonomy', *Philosophical Transactions of the Royal Society B*, 359, published online 29 April (doi: 10.1098/rstb.2003.1447).

Borges, J.L. (1998) *Collected Fictions*, New York: Penguin Books.

Bowker, G. (1994a) *Science on the Run: Information Management and Industrial Geophysics at Schlumberger, 1920–1940*, Cambridge, MA: MIT Press.

Bowker, G. (1994b) 'Information mythology: the world of/as information', in L. Bud-Frierman (ed.) *Information Acumen: The Understanding and Use of Knowledge in Modern Business*, New York: Routledge, pp.231–247.

Bowker, G. (2000) 'Biodiversity data-diversity', *Social Studies of Science*, 30 (5): 643–683.

Bowker, G. (2005) *Memory Practice in the Sciences*, Cambridge, MA: MIT Press.

Bowker, G. and Star, L. (1999) *Sorting Things Out: Classification and its Consequences*, Cambridge, MA: MIT Press.

Bowker, G., Baker, K., Millerand, F. and Ribes, D. (2010) 'Toward information infrastructure studies: ways of knowing in a networked environment', *International Handbook of Internet Research*: 97–117 (doi: 10.1007/978-1-4020-9789-8_5).

Brockington, D. and Duffy, R. (2010) 'Capitalism and conservation: The production and reproduction of biodiversity conservation', *Antipode*, 42 (3): 469–484.

Brockington, D., Duffy, R. and Igoe, J. (2008) *Nature Unbound: The Past, Present and Future of Protected Areas*, London: Earthscan.

Brooke, J.H. (1998) *Reconstructing Nature: The Engagement of Science and Religion*, Oxford: Oxford University Press.

Brower, A. (2006) 'Problems with DNA barcodes for species delimitation: "ten species" of Astrapes fulgerator reassessed (Lepidoptera: Hesperiidae)', *Systematic Biodiversity*, 4: 127–132.

Brower, A. (2010) 'Alleviating the taxonomic impediment of DNA barcoding and setting a bad precedent: names for ten species of "Astraptes fulgerator" (Lepidoptera: Hesperiidae: Eudaminae) with DNA based diagnoses', *Systematics and Biodiversity*, 8 (4): 485–491.

Brown, N. (2003) Hope against hype: accountability in biopasts, presents and futures, *Science Studies*, 16 (2): 3–21.

Brown, N., Rappert, B. and Webster, A. (2000) *Contested Futures: A Sociology of Propsective Techno-science*, Aldershot: Ashgate.

Burke, D. (2004) 'GM food and crops: what went wrong in the UK?' *European Molecular Biology Organisation Reports*, 5: 432–436. Available at: http://www.emboreports. org, accessed 12 March 2013.

Busch, L. (2007) *Universities in the Age of Corporate Science: The UC Berkeley–Novartis Controversy*, Philadephia: Temple University Press.

Cain, A.J. (1962) 'The evolution of taxonomic principles', in Ainsworth, G.C. and Sneath, P.H.A. (eds) *Microbial Classification*, New York: Cambridge University Press, pp.1–13.

Callon, M. (1986) 'Some elements of a sociology of translation: domestication of the scallops and the fishermen of St Brieuc Bay', in Law, J. (ed.) *Power, Action and Belief: A New Sociology of Knowledge?* London: Routledge.

Callon, M. (2004) 'Is science a public good?' *Science, Technology & Human Values*, 19 (4): 395–424.

Cambell-Kelly, M. and Aspray, W. (1996) *Computer: A History of the Information Machine*, New York: Basic Books/HarperCollins.

Cambrosio, A. (2009) 'Section seven: new forms of knowledge production: introduction', in Atkinson, P., Glasner, P. and Lock, M. *Handbook of Genetics and Society: Mapping the New Genomic Era*, London: Routledge.

Castree, N. (2008) 'Neoliberalising nature: processes, effects and evaluations', *Environment and Planning A*, 40 (1): 153–173.

Caterino, M.S., Cho, S. and Sperling, F.A.H. (2000) 'The current state of insect molecular systematics: a thriving Tower of Babel', *Annual Review of Entomology*, 45: 1–54.

Ceruzzi, P.E. (2000) *History of Modern Computing*, Cambridge, MA: MIT Press.

Chase, M.W. *et al.*, (2007) 'A proposal for a standardized protocol to barcode all land plants', *Taxon*, 56, 295–299.

Chow-White, P.A. and García Sancho, M. (2011) 'Global genome databases between biology and computing from the first DNA sequencers to bidirectional shaping and spaces of convergence: interactions', *Science Technology Human Values*, published online 27 February 2011 (doi: 10.1177/0162243910397969).

Claridge, M. (2002) 'Systematic biology and biodiversity', *Antenna*, 26: 86–88.

Clarke, A.E. and Fujimura, J. (eds) (1992 [2002]) *The Right Tools for the Job: At Work in Twentieth-Century Life Sciences*, Princeton, NJ: Princeton University Press.

Clarke, A.E. and Star, S.L. (2008). 'Social worlds/arenas as a theory-methods package', in Hackett, E., Amsterdamska, O., Lynch, M. and Wacjman, J. (eds) *Handbook of Science and Technology Studies*, 2nd edition, Cambridge, MA: MIT Press, pp.113–37.

Connor, S. (2009) 'Michel Serres: the hard and the soft', talk given at the Centre for Modern Studies, University of York, 26 November 2009.

Cook-Degan, R. (1994) *The Gene-Wars: Science, Politics and the Human Genome*, New York: W.W. Norton.

Cooper, Melinda. (2008) *Life as Surplus: Biotechnology and Capitalism in the Neoliberal Era*, Seattle: University of Washington Press.

Costa, F.O. and Carvalho, G.R. (2007) 'The Barcode of Life Initiative: synopsis and prospective societal impacts of DNA barcoding of fish', *Genomics, Society and Policy*, 3 (2): 29–40.

Cox, P. (1990) 'Ethnopharmacology and the search for new drugs', in Chadwick, D.J. and Marsh, J. (eds) *Proceedings of the Ciba Foundation Symposium: Bioactive Compounds from Plants*, The Ciba Foundation, Chichester: Wiley, pp.40–55.

Cox P.A. (2000) 'Will tribal knowledge survive the Millennium?' *Science*, 287 (5450): 44–46.

Crutzen, P.I. and Stoermer, E.F. (2000) 'The "Anthropocene"', *IGBP Newsletter*, 41 (12): 12.

Culp, J.M., Hose, G.C., Armanini, D.G., Dunbar, M.J., Orlofske, J.M., Poff, N.L., Pollard, A.I. and Yates, A.G. (2011) 'Incorporating traits in aquatic biomonitoring to enhance causaldiagnosis and prediction', *Integrated Environmental Assessment Management*, 7: 187–197.

Darling, J.A. and Blum, M.J. (2007) 'DNA-based methods for monitoring invasive species: a review and prospectus', *Biological Invasions*, 9: 751–765.

Darling, J.A. and Mahon, A.R. (2011) 'From molecules to management: adopting DNA-based methods for monitoring biological invasions in aquatic environments', *Environmental Research*, 111: 978–988.

Daston, L. (2004) 'Type specimens and scientific memory', *Critical Enquiry*, 31: 153–82.

Davies, G. (2010) 'Captivating behaviour: mouse models, experimental genetics and reductionist returns in the neurosciences', *The Sociological Review*, 58 (Issue Supplement s1): 53–72.

de Carvalho, D. Neto, D., Brasil, B. and Oliveira, D. (2011) 'DNA barcoding unveils a high rate of mislabeling in a commercial freshwater catfish from Brazil', 22 (S1): 97–105.

Dean, J. (1979) 'Controversy over classification: a case study from the history of botany', in Barnes, B. and Shapin, S. (eds) *Natural Order: Historical Studies of Scientific Culture*, London: Sage Publications.

Deleuze, G. and Guattari, F. (1987) *A Thousand Plateaus: Capitalism and Schizophrenia*, trans. Brian Massumi, Minneapolis: University of Minnesota Press.

Derrida, J. (1988a) 'Signature event context', *Limited Inc*, trans. Samuel Weber and Jeffrey Mehlman, Evanston, IL: Northwestern University Press.

Derrida, J. (1988b) *Of Grammatology*, Baltimore, MD: John Hopkins University Press.

Derrida, J. (1995 [1998]) *Archive Fever: A Freudian Impression*, Chicago, IL: University of Chicago Press. (Page references refer to the 1998 paperback edition.)

DeSalle, R. and V.J. Birstein (1996) 'PCR identification of black caviar', *Nature*, 381: 197–198.

DeSalle, R., Egan, M.G. and Siddall, M. (2005) 'The unholy trinity: taxonomy, species delimitation and DNA barcoding', *Philosophical Transactions of the Royal Society B*, 360: 1905–1916 (doi: 10.1098/rstb.2005.1722).

Dierkes, M. and Von Grote, C. (eds) (2000) *Between Understanding and Trust: The Public, Science and Technology*, London: Routledge.

Dugdale, A. (1999) 'Materiality: juggling sameness and difference', in Law, J. and Hassard, J. (eds) *Actor Network Theory and After*, Oxford: Blackwell.

Durkheim, E. and Mauss, M. (1963 [1903]) *Primitive Classification*, Chicago, IL: University of Chicago Press.

Ebach, M.C. (2011) 'Taxonomy and the DNA barcoding enterprise', *Zootaxa*, 2742: 67–68.

Ebach, M.C. and Holdrege, C. (2005) 'DNA barcoding is no substitute for taxonomy', *Nature*, 434: 697–697 (doi: 10. 1038/434697b).

Ebach, M.C. and de Carvalho M.R. (2010) 'Anti-intellectualism and the DNA barcoding enterprise', *Zoologia*, 27: 165–178.

Eddy, S.R. (1998) 'Profile hidden Markov models', *Bioinformatics*, 14: 755–763.

Edge, D. (1975) 'On the purity of science', in Niblett, W.R. (ed.) *The Sciences, the Humanities and the Technological Threat*, London: University of London Press, pp.42–64.

Eldredge, N. and J. Cracraft. (1980) *Phylogenetic Patterns and the Evolutionary Process. Method and Theory in Comparative Biology*, New York: Columbia University Press.

Elias, M., Hill, R.I., Willmott, K.R., Dasmahapatra, K.K., Brower A.V.Z., Mallet, J. and Jiggins, C.D. (2007) 'Limited performance of DNA barcoding in a diverse community of tropical butterflies', *Proceedings of the Royal Society B*, 274: 2881–2889 (doi: 10.1098/rspb.2007.1035).

Ellis, R. (2009) 'Rethinking the value of biological specimens: laboratories, museums and the Barcoding of Life Initiative', *Museum and Society*, 6 (2): 172–191.

Ellis, R. and Waterton, C. (2004) 'Environmental citizenship in the making: the participation of volunteer naturalists in UK biological recording and biodiversity policy', *Science and Public Policy*, 31 (2): 95–105 (doi: 10.3152/147154304781780055).

Ellis, R. and Waterton, C. (2005) 'Caught between the cartographic and the ethnographic imagination: the whereabouts of amateurs, professionals, and nature in knowing biodiversity', *Environment and Planning D: Society and Space*, 23 (5): 673–693.

Ellis, R., Waterton, C. and Wynne, B. (2010) 'Taxonomy, biodiversity and their publics in 21st century barcoding', *Public Understanding of Science*, 19 (4): 497–512.

Ellul, J. (1964) *The Technological Society*, trans. John Wilkinson, New York: Knopf.

Etzkowitz, H. and Leydesdorff, L. (eds) (1997) *Universities and the Global Knowledge Economy: A Triple Helix of University–Industry–Government Relations*, London: Cassell Academic.

European Commission (2005) *New Perspectives on the Knowledge-based Bio-economy, Conference Report*, Brussels: European Commission.

Ezrahi, Y. (1990) *The Descent of Icarus: Science and the Transformation of Contemporary Democracy*, Cambridge, MA: Harvard University Press.

Fairhead, J. and Leach. M, (1996) *Misreading the African Landscape: Society and Ecology in a Forest–Savanna Mosaic*, London: Cambridge University Press

Fara, P. (2011) *Science: A Four Thousand Year History*, Oxford: Oxford University Press.

Fava, D.S.M. (2009) *Designing Nightmares: Scientific and Artistic Representations of Climate Change as Apocalypse*, PhD Lancaster University.

Felman, S. (2002) *The Scandal of the Speaking Body: Don Juan With J.L. Austin*, Stanford, CA: Stanford University Press.

Finaly, B.J. (2004) 'Protist taxonomy: an ecological perspective', *Philosophical Transactions of the Royal Society B*, 359, published online 29 April (doi: 10.1098/rstb.2003.1450).

Fischer, M.J. (2003) *Emergent Forms of Life and the Anthropological Voice*, Durham, NC: Duke University Press.

Fischer, M.J. (2005) 'Technoscientific infrastructures and emergent forms of life: a commentary', *American Anthropologist*, 107, 1: 55–61.

Folmer, O., Black, M., Hoeh, W., Lutz, R. and Vrijenhoek, R. (1994) 'DNA primers for amplification of mitochondrial cytochrome *c* oxidase subunit I from diverse metazoan invertebrates', *Mol. Mar. Biol. Biotechnol*, 3: 294–299.

Fortun, M. (2001) Mediated speculations in the genomics future markets, *New Genetics and Society*, 20 (2): 139–156.

Fortun, M. (2008) *Promising Genomics, Iceland and deCODE Genetics in a World of Speculation*, Berkeley: University of California Press.

Foucault, M. (1992 [1966]) *The Order of Things: An Archaeology of the Human Sciences*, London: Routledge. Originally published in 1966 in French as *Les Mots et les Choses*, Editions Gallimard.

Franklin, S. (2003) 'Rethinking nature–culture: anthropology and the new genetics', *Anthropological Theory*, 3 (1): 65–85.

Franklin, S. (2005) 'Stem Cells R Us@ emergent life forms and the global biological', in Ong, A. and Collier, S. (eds) *Global Assemblages: Technology, Politics, and Ethics as Anthropological Problems*, Oxford: Blackwell, pp.59–78.

Franklin, S. and Lock, M. (eds) (2003) *Remaking Life and Death: Toward an Anthropology of the Biosciences*, Santa Fe, NM: School of American Research Press; Oxford: James Currey.

Franklin, S. and Lock, M. (2003) 'Animation and cessation: the re-making of life and death', in Franklin, S. and Lock, M. (eds) *Re-Making Life and Death: Toward an Anthropology of the Biosciences*, Santa Fe, NM: School of American Research Press; Oxford: James Currey, pp.3–22.

Franklin, S., Lury, C. and Stacey, J. (2000) *Global Nature, Global Culture*, London, Thousand Oaks, CA: Sage Publications.

Fujimura, J.H. (1987) 'Constructing "do-able" problems in cancer research: articulating alignment', *Social Studies of Science*, 17 (2): 257–293.

Fujimura, J.H. (1988) 'The molecular biological bandwagon in cancer research: where social worlds meet', *Social Problems*, 35: 261–83.

Fujimura, J.H. (1996) *Crafting Science: A Sociohistory of the Quest for the Genetics of Cancer*, Cambridge, MA: Harvard University Press.

Fujimura, J.H. (2003) 'Future imaginaries: genome scientists as socio-cultural entrepreneurs', in Goodman, A., Heath, D. and Lindee, S. (eds) *Genetic Nature/Culture: Anthropology and Science Beyond the Two Culture Divide*, Berkeley: University of California Press, pp.176–199.

García-Sancho, M. (2007a) 'The rise and fall of the idea of genetic information (1948–2006)', *Genomics, Society and Policy*, 2 (3): 16–36.

García-Sancho M. (2007b) 'Mapping and sequencing information: the social context for the genomics revolution', *Endeavour*, 31 (1): 18–23.

García-Sancho M. (2009) 'The perception of an information society and the emergence of the first computerized biological databases', in Matsumoto, A. and Nakano, M. (eds) *Human Genome: Features, Variations and Genetic Disorders*, New York: Nova Publishers, pp.257–276.

García-Sancho, M. (2011) 'From metaphor to practices: the introduction of "information engineers" into the first DNA sequence database', *History and Philosophy of the Life Sciences*, 33: 71–104.

Gaudilliere, J. and Rheinberger, H.J. (eds) (2004) *From Molecular Genetics to Genomics: The Mapping Cultures of Twentieth-Century Genetics*, London: Routledge.

Geoghegan-Quinn, M. (2011) EU Comissioner for Research, Development and Innovation, speech on the EU's 'Innovation Emergency' and the new EU R&D Programme, 2014–2020. Available at: http://europa.eu/rapid/pressReleasesAction. do?reference=SPEECH/11/60&type=HTML, accessed 11 July 2012.

Gibbons, M., Nowotny, H., Schwartzman, S., Cott, S. and Trow, M. (1994) *The New Production of Knowledge: The Dynamics of Science and Research in Contemporary Societies*, London: Sage Publications.

Gilbert, W. (1992). 'A vision of the grail', in Kelves, D.J. and Hood, L. (eds) *The Code of Codes: Scientific and Social Issues in the Human Genome Project*, Cambridge, MA: Harvard University Press, pp.83–97.

Godfray, H.C.J. (2002a) 'How might more systematics be funded?' *Antenna*, 26 (1): 11–17. Available at: http://www.ucl.ac.uk/taxome/godfray02.doc, accessed 12 March 2013.

Godfray, H.C.J. (2002b) 'Challenges for taxonomy – the discipline will have to re-invent itself if it is to survive and flourish', *Nature*, 417 (6884): 17–19.

Godfray, H.C.J. (2002c) 'Towards taxonomy's "glorious revolution"', *Nature*, 420 (6915): 461.

Godfray, H.C.J. and Knapp, S. (2004) 'Taxonomy for the twenty-first century – introduction', *Philosophical Transactions of the Royal Society of London Series B-Biological Sciences*, 359 (14444): 559–569.

Graeber, D. (2008) 'A cosmopolitan and (vernacular) democratic creativity: or there never was a west', in Werbner, P. (ed.) *Anthropology and the New Cosmopolitanism*, New York: Berg.

Greene, S. (2004) 'Indigenous people incorporated? culture as politics, culture as property in pharmaceutical bioprospecting', *Current Anthropology*, 45 (2): 211–237.

Grove, R. (1995) *Green Imperialism*, Cambridge: Cambridge University Press.

Hajibabaei, M., de Waard, J.R., Ivanova, N.V., Ratnasingham, S., Dooh, R.J. Kirk, S.L., Mackie, P.M. and Hebert, P.D.N. (2005) 'Critical factors for assembling a high volume of DNA barcodes', *Philosophical Transactions of the Royal Society B Biological Sciences*, 360: 1959–1967 (doi: 10.1098/rstb.2005.1727).

Hajibabaei, M., Singer, G., Hebert, P. and Hickey, D. (2007) 'DNA barcoding: how it complements taxonomy, molecular phylogenetics and population genetics', *Trends in Genetics*, 23: 167–172.

Hajibabei, M., Shokralla, S., Zhou, X., Singer, G. and Baird, D. (2011) 'Environmental barcoding: a next-generation sequencing approach for biomonitoring applications using River Benthos', *Public Library of Science One*, 6, 4: e17497.

Hamels, J., Gala, L., Dufour, S., Vannuffel, P., Zammatteo, N. and Remacle, J. (2001) 'Consensus PCR and microarray for diagnosis of the genus *Staphylococcus*, species, and methicillin resistance', *BioTechniques*, 31: 1364–1372.

Hanner, R., Becker, S., Ivanova, N. and Steinke, D. (2011) 'FISH-BOL and seafood identification: geographically dispersed case studies reveal systemic market substitution across Canada', *Mitochondrial DNA*, 22 (S1): 106–122.

Haraway, D. (1989) *Primate Visions: Gender, Race and Nature in the World of Modern Science*, New York: Routledge.

Haraway, D. (1991) *Simians, Cyborgs and Women: The Reinvention of Nature*, New York: Routledge.

Haraway, D. (1997) *Modest Witness @ Second Millenium: FemaleMan Meets OncoMouse: Feminism and Technoscience*, New York, London: Routledge.

Haraway, D. (2007) *When Species Meet*, Minneapolis: University of Minnesota Press.

Haraway, D. (2010), 'When species meet: staying with the trouble', *Environment and Planning D – Society and Space*, 28 (1): 53–55.

Harmon, A. (2010), 'Havasupai case highlights risks in DNA research', *The New York Times*, 21 July 2010, p.A5.

Hastrup, K. (1993) 'Hunger and the hardness of facts', *Man*, 28 (4): 727–739.

Hayden, C. (2003a) *When Nature Goes Public: The Making and Unmaking of Bio-prospecting in Mexico*, Princeton, NJ: Princeton University Press.

Hayden, C. (2003b) 'From market to market: bioprospecting's idioms of inclusion', *American Ethnologist*, 30 (3): 359–71.

Hebert P.D.N. and Gregory T.R. (2005) 'The promise of DNA barcoding for taxonomy', *Systematic Biology*, 54: 852–859.

Hebert, P.D.N., Cywinska, A., Ball, S.L. and deWaard, J.R. (2003a) 'Biological identifications through DNA barcodes', *Proceedings of the Royal Society of London Biological Sciences*, 270: 313–322.

Hebert, P.D.N., Penton, E., Burns, J., Janzen, D. and Hallwachs, W. (2004a) 'Ten species in one: DNA barcoding reveals cryptic species in the neotropical skipper butterfly Astraptes fulgerator', *Proceedings of the National Academy of Sciences of the USA*, 101 (41): 14812–14817.

Hebert, P.D.N., Ratsingham, S. and deWaard, J.R. (2003b) 'Barcoding animal life: cytochrome c oxidase sub-unit1 divergences among closely related species', *Proceedings of the Royal Society London S96-S99* (doi: 10.1098/rsbl.2003.0025) (B 270 (Suppl.)).

Hebert, P.D.N., Stoeckle, M.Y., Zemlak, T.S. and Francis, C.M. (2004b) 'Identification of birds through DNA barcodes', *Public Library of Science Biology*, 2: 1657–1663 (doi: 10.1371/journal.pbio.0020312).

Helmreich S. (2008) 'Species of biocapital', *Science as Culture*, 17 (4): 463–478

Helmreich, S. (2009) *Alien Ocean: Anthropological Voyages in Microbial Seas*, Berkeley: University of California Press.

Hennig, W. (1966) *Phylogenetic Systematics*, Urbana: University of Illinois Press.

Hesse, M.B. (1974) *The Structure of Scientific Inference*, Berkeley: University of California Press.

Hine, C. (2006) 'Databases as scientific instruments and their role in the ordering of scientific work', *Social Studies of Science*, 36 (2): 269–298.

Hine, C. (2008) *Systematics as Cyberscience: Computers, Change, and Continuity in Science*, Cambridge, MA: MIT Press.

Hoffmeyer, J. (1996) *Signs of Meaning in the Universe*, Bloomington: Indiana University Press.

Hollingsworth, M., Clark, A., Forrest, L., Richardson, J., Pennington, R.T., Long, D., Cowan, R., Chase, M., Gaudeul, M. and Hollingsworth, P. (2008) 'Selecting barcode loci for plants: evaluation of seven candidate loci wtih species-level sampling in three divergent groups of land plants', *Molecular Ecology Resources*, 9: 439–457.

Hollingsworth, P.M. (2008). 'DNA barcoding plants in biodiversity hot spots: progress and outstanding questions', *Heredity*, 101 (1): 1–2.

Holloway, M. (2006) 'Democratizing taxonomy', *Conservation in Practice*, 7 (2): 14–21.

House of Lords, Select Committee on Science and Technology (1992) *Systematic Biology Research 1st Report*, HL Paper 22–1, London.

House of Lords Select Committee on Science and Technology (2002) *What on Earth? The Threat to the Science Underpinning Conservation*, 3rd Report, HL Paper 118 (i). London.

Hudson, R. and Turelli, M. (2003) 'Stochasticity overrules the 'three-times rule': genetic drift, genetic draft and coalescence times for nuclear loci versus mitochondrial DNA', *Evolution*, 57: 182–190 (doi: 10.1111/j.0014-3820.2003.tb00229.x).

Hull, D. (1988) *Science as a Process: An Evolutionary Account of the Social and Conceptual Development of Science*, Chicago, IL: University of Chicago Press.

Hulme, M. (2009) *Why We Disagree About Climate Change*, London and New York: Cambridge University Press.

Hulme, M. (2011) 'Meet the humanities', *Nature Climate Change*, 1: 177–179 (doi: 10.1038/nclimate11502000).

Humphreys, A.M., Hollingsworth, M.L. and James, K.E. (2005) Report on a visit to Paul Hebert's DNA barcoding laboratory, Guelph Barcode of Life Initiative and the Canadian Barcode of Life Network, University of Guelph, Ontario, Canada, 15–17 June 2005. London: Natural History Museum.

Ingold, T. (2000) *The Perception of the Environment: Essays on Livelihood, Dwelling and Skill*, London: Routledge.

Ivanova, N., De Waard, J. and Hebert, P. (2006) 'An inexpensive automation-friendly protocol for recovering high-quality DNA', *Molecular Ecology Notes*, 6 (4): 998–1002.

Jablonka, E. and Lamb, M. (2005) *Evolution in Four Dimensions: Genetic, Epigenetic, Behavioural and Symbolic Variation in the History of Life*, Cambridge, MA: MIT Press.

Janzen, D. (2004a) 'Setting up tropical biodiversity for conservation through non-damaging use: participation by parataxonomists', *Journal of Applied Ecology*, 41 (1): 181–187.

Janzen, D. (2004b) 'Now is the time', *Philosophical Transactions of the Royal Society, London*, 359: 731–732.

Janzen, D. (2005) 'Use of DNA barcodes to identify flowering plants', *Proceedings of the National Academy of Sciences*, 102: 8369–8374.

Janzen, D.H., Hajibabaei, M., Burns, J.M., Hallwachs, W., Remigio, E. and Hebert, P.D.N. (2005) 'Wedding biodiversity inventory of a large and complex Lepidoptera fauna with DNA barcoding', *Philosophical Transactions of the Royal Society B Biological Sciences*, 360 (1462): 1835–1846.

Jasanoff, S. (2005) *Designs on Nature: Science and Democracy in Europe and the United States*, Princeton, NJ: Princeton University Press.

Jasanoff, S. and Wynne, B. (1998) 'Scientific knowledge and decision making', in Rayner, S. and Malone, E. (eds) *Human Choice & Climate Change*, 4 vols, Columbus, OH: Battelle Press, pp.1–112.

Kamau, E., Fedder, B. and Winter, G. (2010) 'The Nagoya Protocol on access to genetic resources and benefit sharing: what is new and what are the implications for provider and user countries and the scientific community?' *Law, Environment and Development Journal*, 6 (3): 246–267.

Katz, C. (1998) 'Whose nature, whose culture? Private productions of space and the "preservation" of nature', in Braun, B. and Castree, N. (eds) *Remaking Reality: Nature at the Millennium*, London: Routledge, pp.46–63.

Kay, L. (1993) *The Molecular Vision of Life: Caltech, the Rockefeller Foundation and the Rise of the New Biology*, Oxford: Oxford University Press.

Kay, L. (1995) 'Who wrote the book of life? Information and the transformation of molecular biology, 1945–55', *Science in Context*, 8: 609–634.

Kay, L. (2000) *Who Wrote the Book of Life? A History of the Genetic Code*, Stanford, CA: Stanford University Press.

Keepin, W. and Wynne, B. (1984) 'Technical analysis of IIASA energy scenarios', *Nature* 312: 691–695 (20 December) (doi: 10.1038/312691a0).

Keller, E.F. (1995) *Refiguring Life: Metaphors of Twentieth Century Biology*, New York: Columbia University Press.

Keller, E.F. (2000) *The Century of the Gene*, Cambridge, MA: Harvard University Press.

Keller, E.F. (2008) 'Nature and the natural', *BioSocieties*, 3: 117–124 (doi:10.1017/S1745855208006054).

Kellert, S.R. and Wilson, E.O. (1993) *The Biophilia Hypothesis*, Washington, DC: Island Press.

Kenney, M. (1986) *Biotechnology: The University–Industrial Complex*, New Haven: Yale University Press.

Kirby, V. (2011) *Quantum Anthropologies: Life at Large*, Durham and London: Duke University Press.

Knorr Cetina, K. (1999) *Epistemic Cultures: How the Sciences Make Knowledge*, Cambridge, MA: Harvard University Press.

Kohler, F. (2007) 'From DNA taxonomy to barcoding: how a vague idea developed into a biosystematic tool', *Mitt. Mus. Nat.kd. Berl., Zool. Reihe*, 83, (suppl.): 44–51 (doi 10.1002/mmnz.200600025).

Kress, W.J. and Erickson, D.L. (2007) 'A two-locus global DNA barcode for land plants: the coding rbcl gene complements the non-coding trnH-psbA spacer region', *Public Library of Science ONE*, 2: e508.

Kress, W.J. and Erickson D.L. (2008) 'DNA barcodes: genes, genomics, and bioinformatics', *Proceedings of the National Academy of Sciences*, 105 (8): 2761–2762.

Kress, W.J., Wurdack, K.J., Zimmer, E.A., Weigt, L.A. and Janzen, D. (2005) 'Use of DNA barcodes to identify flowering plants', *Proceedings of the National Academy of Sciences*, 102: 8369–8374.

Kuhn, T.S. (1962) *The Structure of Scientific Revolutions*, Chicago, IL: Chicago University Press.

Kwa, C. (2009) 'Maps and the taxonomic style', in Drenthen, M., Keulartz, J. and Proctor, J. (eds) *New Visions of Nature: Complexity and Authenticity*, Dordrecht: Springer, pp.173–177.

Lahaye, R., van der Bank, M., Bogarin, D., Warner, J., Pupulin, F., Gigot, G., Maurin, O., Duthoit, S., Barraclough, T.G., Savolainen, V. (2008) 'From the cover: DNA barcoding the floras of biodiversity hotspots', *Proceedings of the National Academy of Sciences*, 105 (8): 2923–2928.

Larson, B. (2011) *Metaphors for Environmental Sustainability: Redefining Our Relationship with Nature*, New Haven, CT: Yale University Press.

Latour, B. (1992) 'Where are the missing masses? The sociology of a few mundane artifacts', in Bijker, W.E. and Law, J. (eds) *Shaping Technology, Building Society: Studies in Sociotechnical Change*, Cambridge, MA: MIT Press, pp.225–258.

Latour, B. (1993) *We Have Never Been Modern*, Cambridge, MA: Harvard University Press.

Latour, B. (1999) 'Circulating reference: sampling the soil in the Amazon Forest', in Latour, B. *Pandora's Hope: Essays on the Reality of Science Studies*, Cambridge, MA: Harvard University Press, pp.24–79.

Latour, B. (2004) *Politics of Nature*, Cambridge, MA: Harvard University Press.

Latour, B. (2010) 'An attempt at a "compositionist manifesto"', *New Literary History*, 41: 471–490

Law, J. (2002) *Aircraft Stories: Decentering the Object in Technoscience*, Durham and London: Duke University Press.

Law, J. (2004) *After Method: Mess in Social Science Research*, London: Routledge.

Leach, M. and Fairhead, J. (2007) *Vaccine Anxieties*, London: Earthscan.

Ledford, H. (2008) 'Botanical identities: DNA barcoding for plants comes a step closer', *Nature*, 415: 616.

Lennox, J.G. (1985) 'Are Aristotelian species eternal?' in Gotthelf, A. (ed.) *Aristotle on Nature and Living Things: Philosophical and Historical Studies*, Pittsburgh, PA: Mathesis Publications, pp.67–94.

Lenoir, T. (1998) 'Shaping biomedicine as an information science', in Bowden, M.E. Hahn, T.B. and Williams, R.V. (eds) *Proceedings of the 1998 Conference on the History and Heritage of Science Information Systems*, ASIS Monograph Series. Medford, NJ: Information Today, Inc., pp.27–45.

Levidow, L. (2009) 'Making Europe unsafe for agbiotech', in Atkinson, P., Glasner, P. and Lock, M. (eds) *Handbook of Genetics and Society: Mapping the New Genomic Era*, London: Routledge, pp.110–126.

Lewontin, R. (2000) 'Foreword', in Oyama, S. *The Ontogeny of Information: Developmental Systems and Evolution*, Durham, NC: Duke University Press.

Lipscomb, D., Platnick, N. and Wheeler, Q. (2003) 'The intellectual content of taxonomy: a comment on DNA taxonomy', *Trends in Ecology and Evolution*, 18: 65–66.

Little, D. and Stevenson, D.W. (2007) 'A comparison of algorithms for the identification of specimens using DNA barcodes: examples from gymnosperms', *Cladistics*, 23: 1–21.

Lock, M. (2000) 'Deadly disputes: the calculation of meaningful life', in Lock, M., Young, A. and Cambrosio, A. (eds) *Living and Working with the New Medical Technologies*, Cambridge: Cambridge University Press.

Lynch, M. and McNally, R. (2005) 'Chains of custody: visualisation, representation, and accountability in the processing of forensic DNA evidence', *Communication and Cognition*, 38 (3 and 4): 297–318.

Lynch, M., Cole S., McNally, R. and Jordan, K. (2008) *Truth Machine: The Contentious History of DNA Fingerprinting*, Chicago, IL: University of Chicago Press.

Mackenzie, A. (2002) *Transductions: Bodies and Machines at Speed*, New York: Continuum.

Mackenzie, A. (2003) 'Bringing sequences to life: how bioinformatics corporealizes sequence data', *New Genetics and Society*, 22: 315–332.

Mackenzie, A. (2005) 'The performativity of code: software and cultures of circulation', *Theory Culture Society*, 22 (1): 71–92.

Mackenzie, A. (2006) *Cutting Code: Software and Sociality*, New York: Peter Lang.

Mackenzie, A. (2011) 'More parts than elements: how databases multiply', *Environment and Planning D: Society and Space* (doi 10.1068/d6710).

Mackenzie, A., Waterton, C., Busch, L., Ellis, R., Frow, E., McNally, R. and Wynne, B. (2013) 'Bringing standards to life? Three cases in contemporary science', *Science, Technology and Human Values*, 1–22. Doi: 10.1177/0162243912474324.

McKenzie, D. (1989) *Inventing Accuracy: A Historical Sociology of Missile-Guidance Testing*, Cambridge, MA: MIT Press.

McNally, R. and Glasner, P. (2007) 'Survival of the gene? 21st-century visions from genomics, proteomics and the new biology', in Glasner, P., Atkinson, P. and Greenslade, H. (eds) *New Genetics, New Social Formations*, London: Routledge.

McNeil, M. (2005) 'Introduction: postcolonial technoscience', *Science as Culture*, 14 (2): 105–112. Mallet, J. and Willmott, K. (2003) 'Taxonomy: renaissance or Tower of Babel?' *Trends in Ecology and Evolution*, 18: 57–59.

Marcus, G. (1995) 'Ethnography in/of the world system: the emergence of multi-sited ethnography', *Annual Review of Anthropology*, 24: 95–117.

Matsuda, J. (1996) *The Memory of the Modern*, Oxford: Oxford University Press.

Meier, R. (2008) 'DNA Sequences in taxonomy: opportunities and challenges', in Wheeler, Q.D. (ed.) *The New Taxonomy*, Tempe, AZ: Arizona State University.

Meier, R., Shiyang, K., Vaidya, G. and Ng, P.K.L. (2006) 'DNA barcoding and taxonomy in Diptera: a tale of high intraspecific variability and low identification success', *Systematic Biology*, 55: 715–728 (doi: 10.1080/10635150600969864).

Meyer, C.P. and Paulay, G. (2005) 'DNA barcoding: error rates based on comprehensive sampling', *Public Library of Science Biology*, 3: 2229–2238.

Michael, M. (1992) 'Lay discourses of science: science-in-general, science-in-particular, and self', *Science, Technology and Human Values*, 17 (3): 313–333.

Midgley, M. (1992) *Science as Salvation: A Modern Myth and Its Meaning*, London: Routledge.

Miller, S.E. (2007) 'DNA barcoding and the renaissance of taxonomy', *Proceedings of the National Academy of Sciences*, 104 (12): 4775–4776.

Millerand, F. and Bowker, G.C. (2009) 'Metadata standards, trajectories and enactment in the life of an ontology', in Star, S.L. and Lampland, M. (eds) *Formalizing Practices: Reckoning with Standards, Numbers and Models in Science and Everyday Life*, Ithaca, NY: Cornell University Press, pp.149–167.

Mirowski, P. (2011) *Science-Mart: Privatizing American Science*, Cambridge, MA: Harvard University Press.

Mirowski, P. and Mirjam-Sent, E. (eds) (2002) *Science Bought and Sold: Essays in the Economics of Science*, Chicago, IL: University of Chicago Press.

Monaghan, M.T., Balke, M., Pons, J. and Vogler, A.P. (2006) 'Beyond barcodes: complex DNA taxonomy of a south pacific island radiation', *Proceedings of the Royal Society B Biology*, 273: 887–893 (doi: 10.1098/rspb.2005.3391).

Mora, C., Tittensor, D.P., Adl, S., Simpson, A.G.B. and Worm, B. (2011) 'How many species are there on Earth and in the ocean?' *Public Library of Science Biology*, 9 (8): e1001127 (doi: 10.1371/journal.pbio.1001127).

Moritz and Cicero (2004) 'DNA barcoding: promise and pitfalls', *Public Library of Science Biology*, 2 (10): e354 (doi: 10.1371/journal.pbio.0020354).

Murphy, J. and Levidow, L. (2006) *Governing the Transatlantic Conflict Over Agricultural Biotechnology*, London: Routledge.

Muturi, C.N., Ouma, J.O., Malele, I.I., Ngure, R.M., Rutto, J.J. *et al.* (2011) 'Tracking the feeding patterns of Tsetse flies (Glossina Genus) by analysis of bloodmeals using mitochondrial cytochromes genes', *Public Library of Science ONE*, 6 (2): e17284 (doi: 10.1371/journal.pone.0017284).

Nanney, D.L. (1982) 'Genes and phenes in *Tetrahymena*', *Bioscience*, 32: 783–788.

Natural History Museum (2003) *The 2003–2007 Corporate Plan*, London: Natural History Museum.

Nelkin, D. (ed.) (1979) *Controversy: Politics of Technical Decisions*, Beverly Hills, CA: Sage Publications.

Nelkin, D. and Lindee, S. (2004) *The DNA Mystique: The Gene as Cultural Icon*, Michigan: University of Michigan Press.

Newmaster, S., Fazekas, A., Steeves R. and Janovec, J. (2008) 'Testing candidate plant barcode regions in the Myristicaceae', *Molecular Ecology Notes*, 8: 480–490.

Nicholls, H. (2004) 'DNA the barcode of life?' Originally published on behalf of Elsevier by BioMedNet. Now available at: http://www.uoguelph.ca/~phebert/media/BioMedNet%20News%20article.pdf, accessed 12 March 2013.

Nightingale, P. and Martin, M. (2004) 'The myth of the biotech revolution', *Trends in Biotechnology*, 22: 564–569.

Noble, D. (1999) *The Religion of Technology: The Divinity of Man and the Spirit of Invention*, New York: Penguin Books.

Nowotny, H., Scott, P. and Gibbons M. (2001) *Re-thinking Science: Knowledge and the Public in an Age of Uncertainty*, London: Polity.

Nyhart, L.K. (1996) 'Natural history and the "new" biology', in Jardine, N., Secord, J.A. and Spary, E.C. (eds) *Cultures of Natural History*, Cambridge: Cambridge University Press, pp.426–443.

Nzeduru, C.V., Ronca, S. and Wilkinson, M.J. (2012) 'DNA barcoding simplifies environmental risk assessment of genetically modified crops in biodiverse regions', *PLoS ONE*, 7 (5): e35929 (doi: 10.1371/journal.pone.0035929).

Oldham, P. (2004) Submission by The European Community, presented to the Third Meeting – Bangkok Thailand, 14–18 February 2005 – of the UN CBD Ad-Hoc Open Ended Working Group on Access and Benefits Sharing, on Global Status and Trends in

Intellectual Property Rights: Genomics, Proteomics, and Biotechnology. Available at: http://www.cbd.int/doc/meetings/abs/abswg-03/information/abswg-03-inf-04-en.pdf, accessed 9 July 2012.

Oldham, P.D. (2009) *Global Status and Trends in Intellectual Property Claims: Genomics, Proteomics and Biotechnology*. Available at SSRN: http://ssrn.com/abstract=1331514 or http://dx.doi.org/10.2139/ssrn.1331514, (accessed 9 July 2012).

Olds, K. and Thrift, N. (2005) 'Cultures on the brink: reengineering the soul of capitalism – on a global scale', in Ong, A. and Collier, S. (eds) *Global Assemblages: Technology, Politics and Ethics as Anthropological Problems*, Oxford: Blackwell, pp.270–290.

Ong, A. (2005) 'Ecologies of expertise: assembling flows, managing citizenship', in Ong, A. and Collier, S. (eds) *Global Assemblages: Technology, Politics and Ethics as Anthropological Problems*, Oxford: Blackwell, pp.337–353.

Ong, A. and Collier, S.J. (eds) (2005) *Global Assemblages: Technology, Politics and Ethics in Anthropological Problems*, Oxford: Blackwell.

Oren, A. (2004) 'Prokaryote diversity and taxonomy: current status and future challenges', *Philosophical Transactions of the Royal Society B*, 359: 623–638.

Oreskes, N., Shrader-Frechette, K. and Belitz, K. (1994) 'Verification, validation, and confirmation of numerical models in the earth sciences', *Science*, 263 (5147): 641–646.

Oyama, S. (2000) *The Ontogeny of Information: Developmental Systems and Evolution*, Durham, NC: Duke University Press.

Paarlberg. R, (2009) *Starved for Science: How Biotechnology is Being Kept out of Africa*, Cambridge, MA: Harvard University Press.

Pamuk, O. (2009) *The Museum of Innocence: A Novel*, trans. M. Freely. New York: Vintage Books.

Parry, B. (2004) *Trading the Genome: Investigating the Commodification of Bioinformation*, New York: Columbia University Press.

Parry, S. and Dupre, J. (2010) *Nature After the Genome*, Oxford: Blackwell.

Pennisi, E. (2007) 'TAXONOMY: wanted: a barcode for plants', *Science*, 318 (5848): 190–191.

Philip, K. (2005) 'What is a technological author? The pirate function and intellectual property', *Postcolonial Studies*, 8: 199–218.

Pimm, S. (2001) *The World According to Pimm: A Scientist Audits the Earth*, New York: McGraw Hill.

Pimm, S. and Raven. P, (2000) 'Extinction by numbers', *Science*, 233: 2207–2208.

Plotkin, M.J. (2000) *Medicine Quest: In Search of Nature's Healing Secrets*, New York: Viking Press.

Prentice, R. (2005) 'The anatomy of a surgical simulation: the mutual articulation of bodies in and through the machine', *Social Studies of Science*, 35 (6): 837–866.

Puig de la Bellacasa, M. (2011) 'Matters of care in technoscience: assembling neglected things', *Social Studies of Science*, 41 (1): 85–106.

Puillandre, N., Bouchet, P., Boisellier-Dubayle, M.C. *et al.* (2012) 'New taxonomy and old collections: integrating DNA barcoding into the collection curation process', *Molecular Ecology Resources*, 12: 396–402.

Rabinow, P. (1996) *Making PCR: A Story of Biotechnology*, Chicago, IL: University of Chicago Press.

Rabinow, P. (1999) *French DNA: Trouble in Purgatory*, Chicago, IL: The University of Chicago Press.

Ratnasingham, S. and Hebert, P. (2007) 'BOLD: The Barcode of Life Data System', *Molecular Ecology Notes*, 7 (3): 355–364 (doi: 10.1111/j.1471-8286.2006.01678.x).

Reardon, J. (2005) *Race to the Finish: Identity and Governance in an Age of Genomics*, Princeton, NJ: Princeton University Press.

Reardon, J. (2013) 'On the emergence of science and justice', *Science Technology and Human Values*, 38 (2): 176–200.

Reardon, J. and Tallbear, K. (2012), '"Your DNA is our history": genomics, anthropology, and the construction of whiteness as property', *Current Anthropology*, 53 (S5): 233–245.

Renan, E. (1882) '*Qu'est-ce qu'une nation?*' Lecture delivered on 11 March 1882 at the Sorbonne. Republished in 1996 as '*Qu'est-ce qu'une nation?*/What is a nation?' Introduction C. Taylor, trans. W.R. Taylor, Toronto: Tapir Press, pp 21–49.

Richards, T. (1993) *The Imperial Archive: Knowledge and Fantasy of Empire*, London: Verso.

Rheinberger, H.J. and Gaudillière, J.P. (eds) (2004) *From Molecular Genetics to Genomics: The Mapping Cultures of Twentieth Century Genetics*, London, New York: Routledge, pp.95–110.

Ridley, M. (1986) *Evolution and Classification: The Reformation of Cladism*, New York: Longman.

Robertson, M. (2011) 'Measurement and alienation: making a world of ecosystem', *Transactions of the Institute of British Geographers* (doi: 10.1111/j.1475-5661. 2011.00476.x).

Roof, J. (2007) *The Poetics of DNA*, Minneapolis, MN: University of Minnesota Press.

Rose, N. (2007) 'Molecular biopolitics, somatic ethics and the spirit of biocapital', *Social Theory and Health*, 5: 3–29.

Rose, H. and Rose, S. (2012) *Genes, Brains, and Cells: The Promethean Promises of The New Biology*, London and New York: Verso.

Royal Society, The (2003) *Measuring Biodiversity for Conservation*, Policy Document 10/03 (May 2003). Available at: www.royalsoc.ac.uk, accessed 12 March 2013.

Rubinoff, D., Cameron, S. and Will, K. (2006) 'Are plant DNA barcodes a search for the Holy Grail?' *Trends in Ecology & Evolution*, 21: 1–2.

Ruiz Muller, M. (2009) 'CBD ABS principles in the context of barcoded genetic information: the case of iBOL', Proceedings of the Seminar 'Barcoding of Life: Society and Technology Dynamics – Global and National Perspectives', held at the Third International Barcode of Life Conference in Mexico City, Mexico, 9 November 2009.

Schaffer, S. (2005) 'Seeing double: how to make up a phantom body public', in Latour, B. and Weibel, P. (eds) *Making Things Public: Atmospheres of Democracy*, Cambridge, MA: MIT Press, pp.196–202.

Schei, P.J. and Tvedt, J. (2010) '"Genetic resources" in the CBD: the wording, the past, the present and the future', Oslo, Norway: Fridtjof Nansen Institute, Report 4/2010.

Schindel, D. (2010) 'Biology without borders', *Nature*, 467 (7 October): 779–781.

Schindel, D.E. and Miller, S.E. (2005) 'DNA barcoding a useful tool for taxonomists', *Nature*, 435: 17.

Scotland, R. Hughes, C. Bailey, D. and Wortley, A. (2003) 'The big machine and the much-maligned taxonomist. DNA taxonomy and the web', *Systematics and Biodiversity*, 1 (2): 139–143.

Seberg, O., Humphries, C.J., Knapp, S., Stevenson, D.W., Petersen. G., *et al.* (2003) 'Shortcuts in systematics? A commentary on DNA-based taxonomy', *Trends in Ecology and Evolution*, 18: 63–65.

Serres, M. (1994) *Atlas*, Lisbon, Portugal: Piaget.

Serres, M. (2001) *Hominescence*, Paris: Le Pommier.

Serres, M. (2009) *Ecrivains, savants et philosophes font le tour du monde*, Paris: Le Pommier.

Shackley, S. and Wynne, B. (1996) 'Representing uncertainty in global climate change science for policy: boundary-ordering devices and authority', *Science, Technology and Human Values*, 21 (3): 275–302.

Shapin, S. (1994) *A Social History of Truth: Civility and Science in Seventeenth Century England*, Chicago, IL: Chicago University Press.

Shapin, S. and Schaffer, S. (1985) *Leviathan and the Air Pump: Hobbes, Boyle and the Experimental Life*, Princeton, CA: Princeton University Press.

Shorrett P., Rabinow P. and Billings, P.R. (2003) 'The changing norms of the life sciences', *Nature Biotechnology*, 21: 123–25.

Sperling, F. (2003) 'DNA barcoding: deus ex machina', *Newsletter of the Biological Survey of Canada* (Terrestrial Arthropods), 22: 50–53.

Sowa, J. (2000) *Knowledge Representation: Logical, Philosophical, and Computational Foundations*, Pacific Grove, CA: Brooks/Cole.

Stemerding, D. (1991) *Plants, Animals and Formulae: Natural History in the Light of Latour's Science in Action and Foucault's the Order of Things*, Enschede, The Netherlands: University of Twente.

Stengel, K., Taylor, J., Waterton, C. and Wynne, B. (2009) 'Plant Sciences and the public good', *Science, Technology & Human Values*, 34 (3): 289–312

Stengers. I (2010) *Cosmopolitics I*, trans. Robert Bononno, Minneapolis: University of Minnesota Press.

Stengers. I (2011) *Cosmopolitics II*, trans. Robert Bononno, Minneapolis: University of Minnesota Press.

Stoekl, A. (2007b) 'Excess and depletion: bataille's suprisingly ethical model of expenditure', in Winnubst, S. (ed.) *Reading Bataille Now*, Bloomington: Indiana University Press, pp.252–82.

Stoeckle, M. (2003) 'Taxonomy, DNA and the barcode of life', *BioScience*, 53: 2–3.

Stoekle, M. and Hebert, P. (2008) 'Barcode of life', *Scientific American*, 299: 82–88.

Stotz, K., Griffiths, R. and Knight. P. (2004) 'How biologists conceptualize genes: an empirical study', *Studies in the History and Philosophy of Biological and Biomedical Sciences*, 35: 647–673.

Strasser, B.J. (2008) 'GenBank: natural history in the 21st century?' *Science*, 322: 537–538.

Strasser, B.J. (2011) 'The experimenter's museum: genbank, natural history, and the moral economies of biomedicine', *Isis*, 102, 1: 60–96.

Strathern, M. (2002) *Reproducing the Future: Anthropology, Kinship and the New Reproductive Technologies*, New York: Routledge.

Strathern, M. (2004) *Commons and Borderlands. Working Papers On Interdisciplinarity, Accountability and the Flow of Knowledge*, Oxfordshire: Sean Kingston Publishing.

Suchman, L. (1994) 'Working relations of technology production and use', *Computer Supported Cooperative Work (CSCW)*, 2: 21–39.

Suchman, L. (2000) 'Organizing alignment: a case of bridge building', *Organization*, 7 (2): 311–327.

Suchman, L. (2002) 'Practice based design of information systems: notes from the hyperdeveloped world', *The Information Society: An International Journal*, 18 (2): 139–144.

Suchman, L. (2007) *Human–Machine Reconfigurations: Plans and Situated Actions*, Cambridge: Cambridge University Press.

Suchman, L. and Bishop, L. (2000) 'Problematizing "innovation" as a critical project', *Technology Analysis & Strategic Management*, 12 (3): 327–333.

Sullivan, S. (2010) '"Ecosystem service commodities" – a new imperial ecology? implications for animist immanent ecologies, with Deleuze and Guattari', *New Formations* (doi: 10.3898/NEWF.69.06.2010).

Sunder Rajan, K. (2006) *Biocapital: The Constitution of Postgenomic Life*, Durham, NC and London: Duke University Press.

Swyngedouw, E. (2010) 'Apocalypse forever? Post-political populism and the spectre of climate change', *Theory, Culture & Society*, 27 (2–3): 213–232.

Szerszynski, B. (2005) *Nature, Technology and the Sacred*, Oxford: Blackwell.

Taussig, M. (2004) *My Cocaine Museum*, Chicago, IL: Chicago University Press

Taussig, M. (2006) *Walter Benjamin's Grave*, Chicago and London: Chicago University Press.

Tautz, D., Arctander, P., Minelli, A., Thomas, R.H. and Vogler, A. (2002) 'DNA points the way ahead in taxonomy', *Nature*, 418: 479.

Tautz, D., Arctander, P., Minelli, A., Thomas, R.H. And Vogler, A. (2003) 'A plea for DNA taxonomy', *Trends in Ecology and Evolution*, 18 (2): 70–74.

Ten Kate, K. and Laird, S. (2000) 'Biodiversity and business: coming to terms with the "grand bargain"', *International Affairs*, 76: 241–264.

Thacker, E. (2005) *The Global Genome: Biotechnology, Politics and Culture*, Cambridge, MA: MIT Press.

Thévenot, L. (2009) 'Governing life by standards: a view from engagements', *Social Studies of Science*, 39 (5): 793–813.

Thomas, K. (1984) *Man and the Natural World: Changing Attitudes in England 1500–1800*, London: Penguin.

Thompson, C. (2005) *Making Parents: The Ontological Choreography of Reproductive Technologies*, Cambridge, MA: MIT Press.

Toulmin, S. (1982) *Return to Cosmology Postmodern Science and the Theology of Nature*, Berkeley: University of California Press.

Toulmin, S. (1990) *Cosmopolis: The Hidden Agenda of Modernity*, Chicago, IL: University of Chicago Press.

Traweek, S. (1988) *Beamtimes and Lifetimes: The World of High Energy Physics*, Cambridge, MA: Harvard University Press.

Tsing, A. (2005) *Friction: An Ethnography of Global Connection*, Princeton, NJ: Princeton University Press.

Turnhout, E. and Boonman-Berson, S. (2011) 'Databases, scaling practices, and the globalization of biodiversity', *Ecology and Society*, 16 (1): 35. Available at: http://www.ecologyandsociety.org/vol16/iss1/art35 (accessed 12 March 2013).

Turnhout, E., Bloomfield, B., Hulme, M., Vogel, J. and Wynne, B. (2012) 'Listen to the voices of experience', *Nature*, 488: 454–455.

Turnbull, D. (2003) 'Assemblages and diversity: working with incomensurability: emergent knowledge, narrativity, performativity, mobility and synergy', paper presented at AHHPSSS, Melbourne.

Vernooy, R., Haribabu, E., Ruiz Muller, M., Vogel, J.H. and Hebert, P.D.N. (2010) 'Barcoding life to conserve biological diversity: beyond the taxonomic imperative', *Public Library of Science Biology*, 8 (7): e1000417 (doi: 10.1371/journal.pbio.1000417).

Verran, H. (2001) *Science and an African Logic*, Chicago, IL: Chicago University Press.

Vijayan, K. and Tsou, C.C. (2010) 'DNA barcoding in plants: taxonomy in a new Perspective', *Current Science*, 99 (11): 1530–1541.

Vogel, J.H. (2009) 'iBOL as an enabler of ABS and ABS as an enabler of iBOL', proceedings of the seminar 'Barcoding of Life: Society and Technology Dynamics – Global and National Perspectives', held at the Third International Barcode of Life Conference in Mexico City, Mexico, 9 November 2009.

Vogler, A. and Monahan, M. (2006) 'Recent advances in DNA taxonomy', *Journal of Zoological Systematics and Evolutionary Research*, 45 (1): 1–10 (doi: 10.1111/j.1439-0469.2006.00384.x).

Ward, R.D., Zemlak, T.S., Innes, B.H., Last, P.R. and Hebert, P.D.N. (2005) 'Barcoding Australia's fish species', *Philosophical Transactions of the Royal Society B Biology*, 360: 1847–1857.

Webster, A. (1989) 'Privatisation of public sector research: the case of a plant breeding institute', *Science and Public Policy*, 16 (4): 224–32.

Waterton, C. (2002) 'From field to fantasy: classifying nature, constructing Europe', *Social Studies of Science*, 32 (2): 177–204.

Waterton, C. (2010a) 'Barcoding nature: strategic naturalisation as innovatory practice in the genomic order of things', *The Sociological Review*, 58 (Issue Supplement s1): 152–171.

Waterton, C. (2010b) 'Experimenting with the archive: STS-ers as analysts and co-constructors of databases and other archives', *Science Technology and Human Values*, 35: 645–676. Originally published online 26 February 2010 (doi: 10.1177/0162243909340265).

Waterton, C. and Wynne, B. (1996) 'Building the European Union: science and the cultural dimensions of environmental policy', *Journal of European Public Policy*, 3 (3): 421–440.

Weart, S. (1988) *Nuclear Fear: A History of Images*, Cambridge, MA: Harvard University Press.

Werbner, P. (ed.) (2008) *Anthropology and the New Cosmopolitanism*, New York: Berg.

Wheeler, Q. (2004) 'Taxonomic triage and the poverty of phylogeny', *Philosophical Transactions of the Royal Society B, Biological Sciences*, 359 (1444): 571–583.

Wheeler, Q. (2008) 'Introduction: toward the new taxonomy', in Wheeler, Q. (ed.) *The New Taxonomy*, The Systematics Association Special Volume Series 76, New York: CRC Press, pp.1–18.

Wheeler, Q. and Cracraft, J. (1996) 'Taxonomic preparedness: are we ready to meet the biodiversity challenge?' in Wilson, E.O., Kudla-Reaka, M. and Wilson, D. (eds) *Biodiversity II: Understanding and Protecting our Biological Resource*, Washington, DC: National Academy of Sciences Press, pp.435–446.

Wheeler, Q. and Meier, R. (eds) (2000) *Species Concepts and Phylogenetic Theory*, New York: Columbia University Press.

Wheeler, Q.D., Knapp, S., Stevenson, D.W., Stevenson, J. Blum, S.D. *et al.* (2012) 'Mapping the biosphere: exploring species to understand the origin, organization and sustainability of biodiversity', *Systematics and Biodiversity*, 10 (1): 1–20.

Wiesmer, M. and Fiedler, K. (2007) 'Does the DNA barcoding gap exist?—a case study in blue butterflies (Lepidoptera: Lycaenidae)', *Fronteirs in Zoology*, 4 (8) (doi: 10.1186/1742-9994-4-8).

Will, K.W. and Rubinoff, D. (2004) 'Myth of the molecule: DNA barcodes for species cannot replace morphology for identification and classification', *Cladistics*, 20: 47–55 (doi: 10.1111/j.1096-0031.2003.00008.x).

Will, K.W., Mishler, B.D. and Wheeler, Q.D. (2005) 'The perils of DNA barcoding and the need for integrative taxonomy', *Sytematic Biology*, 54 (5): 844–851 (doi: 10.1080/10635150500354878).

Wilson, E.O. (1984) *Biophilia: The Human Bond with Other Species*, Cambridge, MA: Harvard University Press.

Wilson, E.O. (1985) 'The biodiversity crisis: a challenge to science', *Issues in Science and Technology*, 2 (1): 20–29.

Wilson, E.O (1991) 'Arousing biophilia: a conversation with E.O. Wilson', *Orion*, 10 (Winter): 9–15.

Wilson, E.O. (1992) *The Diversity of Life*, Cambridge, MA: Harvard University Press.

Wilson, R.A. (ed.) (1999) *Species: New Interdisciplinary Essays*, Cambridge, MA: MIT Press.

Wolfe, K.H., Li, W.H. and Sharp, P.M. (2007) 'Rates of nucleotide substitution vary greatly among plant mitochondrial, chloroplast and nuclear DNA', *Proceedings of the National Academy of Sciences USA*, 1987 (84): 9054–9058.

Wynne, B. (1989) 'Establishing the rules of laws: constructing expert authority', in R. Smith and B. Wynne (eds) *Expert Evidence: Interpreting Science in The Law*, London and New York: Routledge, pp.31–35.

Wynne, B. (1993) 'Public uptake of science: a case for institutional reflexivity', *Public Understanding of Science*, 2 (4): 321–337 (doi: 10.1088/0963-6625/2/4/003).

Wynne, B. (2005) 'Reflexing complexity: post-genomic knowledge and reductionist returns in public science', *Theory, Culture and Society*, 22 (5): 67–94.

Wynne, B. (2006) 'Public engagement as a means of restoring public trust in science: hitting the notes, but missing the music?' *Public Health Genomics*, 9 (3): 211–220.

Wynne, B. (2010) 'Strange weather, again: climate science as political art', *Theory, Culture and Society*, 27 (2–3): 289–305.

Yusoff, K. (2010) 'Biopolitical economies and the political aesthetics of climate change', *Theory, Culture and Society*, 27 (2–3): 73–99.

Yussof, K. (2011) 'Aesthetics of loss: biodiversity, banal violence and biotic subjects', *Transactions of the Institute of British Geographers* (doi: 10.1111/j.1475-5661.2011.00486.x).

Zebich-Knos, M. (1997) 'Preserving biodiversity in Costa Rica: the case of the Merck-INBio agreement', *The Journal of Environmental Development*, 6 (2): 180–186.

Zhang, D.-X. and Hewitt, G.M. (1997) 'Assessment of the universality and utility of a set of conserved mitochondrial primers in insects', *Insect Molecular Biology*, 6: 143–150.

Index